T0231811

EXPERIMENTAL EVIDENCE AND THEORETICAL APPROACHES
IN UNSATURATED SOILS

PROCEEDINGS OF AN INTERNATIONAL WORKSHOP ON UNSATURATED SOILS
TRENTO/ITALY/10 – 12 APRIL 2000

Experimental Evidence and Theoretical Approaches in Unsaturated Soils

Edited by

A. Tarantino
Università degli Studi di Trento, Italy

C. Mancuso
Università degli Studi di Napoli Federico II, Italy

Taylor & Francis
Taylor & Francis Group

LONDON AND NEW YORK

Organised by the Department of Structural and Mechanical Engineering of the University of Trento, Italy, with the support of the Department of Geotechnical Engineering of the University of Naples Federico II, Italy, under the auspices of the Technical Committee on Unsaturated Soils (TC6) of the International Society of Soil Mechanics and Geotechnical Engineering (ISSMGE) and the Italian Geotechnical Association (AGI).

The texts of the various papers in this volume were set individually by typists under the supervision of each of the authors concerned.

Published by Taylor & Francis
2 Park Square, Milton Park, Abingdon, Oxon, OX14 4RN
270 Madison Ave, New York NY 10016

Transferred to Digital Printing 2007

ISBN 90 5809 186 4

Publisher's Note
The publisher has gone to great lengths to ensure the quality of this reprint but points out that some imperfections in the original may be apparent

Experimental Evidence and Theoretical Approaches in Unsaturated Soils, Tarantino & Mancuso (eds)
© 2000 Taylor & Francis, ISBN 90 5809 186 4

Table of contents

Miscellaneous

Experimental Evidence and Theoretical Approaches in Unsaturated Soils, Tarantino & Mancuso (eds)
© 2000 Taylor & Francis, ISBN 90 5809 186 4

Preface

We happen to be involved in research. But research does not always lead to experimental results and numerical simulations that are in good agreement with each other. Research is also studded with errors, results that are difficult to interpret and theories that go nowhere. Whether working in a laboratory, struggling with apparatus that sometimes makes fun of us, or using a computer that refuses to co-operate and returns nothing but error messages, we understood one important thing. No research can be carried out alone. We need to share ideas, troubles, and doubts. That was the essence of the workshop that took place in Trento on 10 – 12 April 2000.

In line with the aim of the workshop, sufficient time was allowed for each speaker to give a detailed presentation. Each presentation was then followed by an equal amount of time that was given to the audience for discussion. In this respect, we wish to thank the discussion leaders of the three sections of the workshop, who led the long debate following each presentation. As pointed out above, research needs to be fed by continuous discussion. Thus, we also wish to thank the more experienced researchers, often co-authors of the papers presented in this book, for giving us the opportunity to think further.

The works published in this book have been revised following comments, discussion and suggestions from the audience and post conference referee. Authors were invited not to be afraid of presenting works still underway, highlighting limitations and blackpoints of their experimental and theoretical research. We hope this will be of help to other researchers. At the same time, we are aware that the reader of this book will expect to find some advances in unsaturated soil mechanics. We are confident her or his expectations will not be disappointed.

Alessandro Tarantino & Claudio Mancuso

Acknowledgements

Many contributed to the success of the workshop. Among them, we wish to thank Professor Luigi Mongiovì for his continuous support and advice, and Dr Elisabetta Nones of the Conference Office of the University of Trento, who undertook a significant part of the organisation.

Special thanks go to the discussion leaders in the workshop:
Professor S.Aversa, Dr G.Bosco, Professor P.Delage, Dr C.Mancuso, Professor R.Nova, Dr V.Sivakumar.

Last but not least, we wish to thank all the speakers in the workshop: Mauricio Barrera, Marc Buisson, Alessandra Di Mariano, Françoise Geiser, Cristina Jommi, Duilio Marcial, Marco Nicotera, Enrique Romero, John Shevelan, Roberto Vassallo, Jean Vaunat. To you all, *thank you*.

Experimental evidence

Experimental Evidence and Theoretical Approaches in Unsaturated Soils, Tarantino & Mancuso (eds)
© *2000 Taylor & Francis, ISBN 90 5809 186 4*

Experimental investigations on the stress variables governing unsaturated soil behaviour at medium to high degrees of saturation

A. Tarantino & L. Mongiovì
Dipartimento di Ingegneria Meccanica e Strutturale, Università degli Studi di Trento, Italy

ABSTRACT: This paper presents the results from an experimental study on the stress variables governing unsaturated soil behaviour. At the first stage of the study, the case of negative pore pressures and continuous air phase was addressed. A new apparatus was designed and implemented for this purpose. Several tests were performed to verify the common assumption that total stress, pore-air pressure, and pore-water pressure can be combined in two effective stress fields. At the second stage of the study, the case of occluded air was examined. With the operational apparatus, a series of tests was conducted to examine the validity of the axis translation technique at high degrees of saturation. Finally, an osmotically controlled oedometer was set up to investigate whether the effective stress principle for saturated soils is applicable to unsaturated soils at high degrees of saturation.

1 INTRODUCTION

The debate on the unsaturated soil stress state variables has generally focused on the choice of a single stress variable or two independent stress variables approach. However, there are other aspects that need to be investigated at a most fundamental level.

The majority of experimental data on unsaturated soil behaviour has been obtained using the axis translation technique. Its validity lays on the assumption that total stress, pore-air pressure, and pore-water pressure can be reduced to two effective stress fields (e.g., net stress and matrix suction). However, little experimental evidence supports this assumption. It is then fundamental to understand whether axis translation data can be extrapolated to unsaturated soils under atmospheric conditions in the field.

The stress variable problem can be viewed from two different standpoints. At a first level, it is necessary to define the stress variables that make it possible to investigate *in situ* soils throughout laboratory tests. These variables will be later referred to as *effective stresses* for unsaturated soils. At a second level, a stress variable approach has to be chosen to model laboratory data obtained using axis translation. In general, there is no approach that can be discarded a priori and no reason to reject stress variables that incorporate a soil parameter. These variables, however, will not be referred to as effective stresses but merely *stress variables*. Limitations and advantages of the different stress variable approaches have been discussed by Geiser et al. (2000) and Jommi (2000). In this paper, an experimental study carried out to investigate the effective stresses for unsaturated soils is presented.

It has been generally assumed that the total stress σ_{ij}, the pore water pressure u_w, and the pore air pressure u_a can be reduced to two effective stresses, chosen among

$$(\sigma_{ij} - u_a\delta_{ij}), (\sigma_{ij} - u_w\delta_{ij}), (u_a - u_w)\,\delta_{ij} \tag{1}$$

For the case where the air phase is continuous in the interstices, this assumption is supported by the results of only one study (Bishop & Donald 1961). No additional data seem to be available in the case of negative pore water pressure, despite its relevance in engineering practice.

For the case where air phase is in the form of cavity within the pore water, the validity of the axis translation technique is also controversial. Null tests performed by Fredlund & Morgenstern (1977) would confirm that σ_{ij}, u_w, and u_a could be combined in two effective stresses, even when the air phase is occluded. On the other hand, it has been suggested that axis translation technique is no longer valid when air phase is occluded (Bocking & Fredlund 1980). Accordingly, three effective stresses instead of two would be required for soils in this unsaturated state. This is an important point that needs to be fully addressed, since much of the experimental data presented in the literature have been obtained testing samples at high degrees of saturation while using the axis translation technique.

Stemming from the above, an experimental programme has been underway at the Geotechnical Laboratory of the University of Trento to investigate the effective stresses for unsaturated soils. At the first stage of the study, the case of the negative pore pressures and the continuous air phase was addressed. To this end, a new apparatus was designed and implemented. In this paper, the new apparatus is briefly described and results from the first set of samples are summarised. As a second step of the experimental study, the case of occluded air was examined. Using the new apparatus, a series of tests was conducted to examine the validity of the axis translation at high degrees of saturation. This portion of the experimental programme is still in progress, and preliminary results are presented.

To complete the investigation in the case of occluded air phase, another problem was examined. During the desaturation process, two effective stresses take the place of the single effective stress associated with the saturated state. In general, it is assumed that unsaturated soil behaviour is still governed by the single saturated effective stress σ-u_w, provided that matrix suction does not exceed the air entry value. No experimental data, however, seem to be available regarding this transition from one to two variables. To investigate this problem an osmotically controlled oedometer equipped with IC tensiometers for direct measurement of matrix suction was set up. The equipment and some preliminary results are also presented in this paper.

2 EFFECTIVE STRESSES FOR UNSATURATED SOILS

In the development of a mechanistic framework for unsaturated soils, the Bishop's stress variable (Bishop 1959):

$$\sigma' = (\sigma - u_a) - \chi \, (u_a - u_w) \qquad (2)$$

has been named "effective stress", in an attempt to extend a well-established concept for saturated soils to unsaturated soils. On the other hand, any pair of tensors in (1) has been termed "stress state variables" to clearly differentiate the two stress variables approach from that based on a single one. Notwithstanding these reasons, the authors believe that the stress tensors in (1) should be regarded as effective stresses for unsaturated soils, whereas the stress defined by Equation 2 (Bishop's stress) should be simply referred to as a stress variable.

As introduced by Terzaghi (1936), the concept of effective stress is bounded to the notion of measurable effects. Being a macroscopic concept, it can be proven or disproved only on the basis of experimental evidence and not of theoretical models at a microscopic level (e.g. Pietruszczak & Pande 1996).

In saturated soils, experimental results have indicated that measurable effects are controlled by one single variable σ-u_w, which has therefore been named effective. The adjective effective is not synonym of single variable, but simply refers to measurable effects.

In unsaturated soils, three stress variables can be generally defined, namely the total stress σ, the pore-water pressure u_w, and the pore-air pressure u_a. As for the case of saturated soils, where two stress variables can be reduced to a single one, it is also possible that the three variables in unsaturated soils be reduced to two. Provided all the measurable effects are exclusively due to changes in these two variables, they should be named effective stresses. Alonso et al. (1987) are presumably on the same line of thinking as they also refer to net stress and matrix suction as effective stresses for unsaturated soils.

On the contrary, it is not appropriate the use of the term effective stress when referring to the Bishop's stress or similar equations. This is simply because no experimental evidence has shown that the measurable effects are exclusively due to changes of one single variable. In some spe-

4

cific problems, the single variable approach may be conveniently and successfully used to predict soil behaviour (Khalili & Kabbaz 1998, Geiser 2000, Khalili 2000). In some cases it is convenient to force unsaturated soils into the saturated soil framework because of the availability of numerical models based on a single variable (Bolzon & Schrefler 1995). However, it is fair to recognise that soils are generally unsaturated, with the saturated state being only a peculiar case. Therefore, it is the mechanics of saturated soil behaviour that should be accommodated into the framework of unsaturated soil behaviour, and not vice versa.

3 EXPERIMENTAL APPARATUSES

Two experimental apparatuses were used in this study. The first device was conceived at the University of Trento, and was named Wait & Pray (W&P) because of the long time required for complete stress equalisation. This was initially believed to be a significant limitation of the experimental procedure devised. The second device was an osmotically controlled oedometer purchased from the Imperial College of London, and has been subsequently modified to fulfil the needs of this research programme.

3.1 The W&P apparatus

Previous experimental studies on the effective stresses for unsaturated soils have been carried out using the axis translation technique, hence maintaining the pore water pressure in the positive range (Bishop & Donald 1961, Fredlund & Morgenstern 1977). One of the main objectives of the research programme presented in this paper was to extend this practice to the negative range of the pore water pressure. The introduction of a tensiometer capable of measuring water tension up to 1500 kPa (Ridley & Burland 1993) made this goal virtually possible.

The approach followed by Bishop & Donald (1961) and Fredlund & Morgenstern (1977) was deemed difficult to implement in the negative range of u_w. Thus a different experimental procedure was devised herein and, accordingly, a new apparatus was designed and implemented (Mongiovì & Tarantino 1998).

In the developed apparatus, samples are maintained under constant strain state and water content. Keeping the pore water pressure in the negative range, changes in net stress (σ-u_a) and matrix suction (u_a-u_w) are measured in response to applied changes of pore air pressure. If the three stress variables σ, u_w, and u_a can be combined in two effective stresses, the differences σ-u_a and u_a-u_w should remain unchanged regardless of any variation of u_a.

The W&P apparatus is briefly described herein. More details can be found in Tarantino (1998). It consists of a cell, equipped with measurement devices, an air tank, and a pressure chamber (Fig.1). The cell was designed to ensure a constant state of strain of the unsaturated specimen. The specimen, having a diameter equal to the inner diameter of the cell, is first compressed into the cell by a steel cap. Then, to prevent any subsequent dilation, the cap is tightened to the cell by means of bolts. The specimen is 120 mm in diameter, and its height may be varied from 18 to 20 mm.

The tank, which supplies the air to increase the pore air pressure, was designed to ensure constant water content of the unsaturated specimen. It is equipped with an impermeable diaphragm that seals the mass of air that flows inside or outside the soil through three porous discs incorporated into the base of the cell. The impermeable membrane makes it possible to prevent soil water evaporative losses, then ensuring constant water content of the specimen throughout the test.

Although the steel cell and the steel cap are virtually rigid when compared to the soil specimen, some dilation of the metal structure might still occur when the air pressure inside the sample is increased. To prevent this potential dilation, the air pressure outside the cell is maintained at the same value as the air pressure inside the cell. For this purpose, a pressure chamber is mounted over the cell and connected to the air supply system in parallel with the lower compartment of the air tank. Valve G in Figure 1 allows venting or isolating the air tank from the pressure chamber.

The net stress σ-u_a is measured locally by means of four flush diaphragm transducers. Two are positioned centrally at the top and bottom of the cell, while the other two are miniature

5

transducers placed opposite to each other on the inner lateral surface of the cell. All transducers are vented to the pressure chamber, with the back of the measurement diaphragm being subjected to the same air pressure. When the air pressure u_a inside the specimen equals the pressure in the chamber, the normal net stress σ-u_a is directly measured.

The matrix suction is measured with one tensiometer (Ridley & Burland 1996) positioned at the top of the sample and at the furthest point from the porous discs. This measuring device is vented to the pressure chamber through its electric cable. When the air pressure u_a inside the specimen equals the pressure in the chamber, the matrix suction u_a–u_w is directly measured.

The air pressure in the tank and in the chamber is measured by absolute recessed diaphragm pressure transducers.

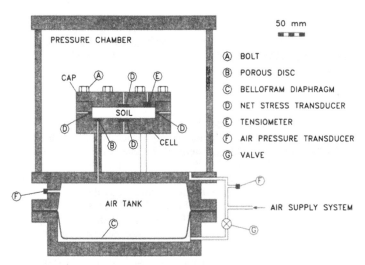

Figure 1. Schematic layout of the W&P apparatus.

3.2 The osmotically controlled oedometer

The validity of the axis translation technique is uncertain when the air phase is occluded within the pore water. That is why the osmotic technique (Delage et al. 1987, Cui & Delage 1996) was chosen to investigate the transition from saturation to the state where the air phase becomes continuous in the interstices.

In the osmotic technique, the negative pore-water pressure is controlled by a semi-permeable membrane in contact with the soil specimen along with a PEG (polyethylene glycol) solution circulating on the other side of the membrane. If no PEG molecules from the solution cross the membrane, soil matrix suction equals the solution's osmotic pressure. However, a certain amount of PEG molecules generally passes in the soil water and the generated osmotic pressure is unknown. As a result, negative pore water pressure needs to be directly measured using tensiometers (Dineen & Burland 1995). At the present stage of the study presented in this paper, it was deemed to be troublesome incorporating one or more tensiometers in an osmotic triaxial apparatus. Thus a simpler apparatus was selected to carry out the investigation.

The oedometer is shown in Figure 2. The osmotic solution is spread underneath the specimen by a nylon woven mesh, according to the original design (Dineen & Burland 1995). To avoid penetration of the specimen into the mesh, a bronze porous disc was then inserted between the semi-permeable membrane and the mesh. This porous element does not slow down the water exchange between the specimen and the osmotic solution (Tarantino & Mongiovì 1999).

The osmotic solution was contained in a closed bottle that rested on a balance. The bottle had two openings where two rubber stoppers had been pushed in. The tubes of the peristaltic pump were therefore forced through a hole drilled into each stoppers so as to ensure air-tightness. The tubes were then firmly secured using a clamp to avoid swaying which could affect the weight measurement. It was verified that the flexible tubes did not prevent the free movement of the balance plate, thus allowing an accurate measurement of the change in the solution mass.

Monitoring the weight of the bottle makes it possible to trace changes in the specimen water content. To avoid dilution underneath the sample, where water exchange occurs, the solution is circulated with a peristaltic pump. PEG with 20000 molecular weight was used to prepare the osmotic solution. Two sheets of Spectrum No. 4 (MWCO 14000), put one on top of the other, forms the semi-permeable interface at the base of the unsaturated soil sample. This arrangement was chosen based on a previous study carried out on three different membranes and using either a single or double sheet interface (Tarantino & Mongiovì 2000).

The negative pore water pressure is measured by two IC tensiometers installed through the top cap. The annular space between the tensiometer and the hole in the cap is sealed with an O-ring to prevent water evaporation from the soil surface where suction is measured. The tensiometer is secured to the cap by means of three bolts. Also, an elastomer membrane is placed over the annular gap between the cap and the inner oedometer ring to prevent soil water evaporation. The membrane attachment was designed to minimise this gap.

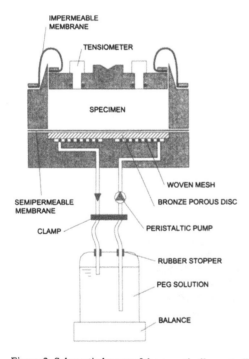

Figure 2. Schematic layout of the osmotically controlled oedometer (Tarantino & Mongiovì 2000).

4 TEST MATERIAL AND SPECIMEN PREPARATION

Kaolin samples ($w_l = 0.54$, and $w_p = 0.32$) consolidated from slurry were used throughout this experimental study. Samples were consolidated at a vertical stress of 100 kPa for use in the W&P, while samples for oedometer testing were consolidated to 500 kPa. Once removed from the consolidometer, all samples were air dried to a given degree of saturation estimated by sample weighing. Thereafter, they were sealed in two polythene bags, placed in an airtight plastic box, and stored for a period of at least one week to allow moisture equalisation.

5 EXPERIMENTAL PROCEDURE

5.1 *Null tests*

The unsaturated specimen was trimmed to 120 mm diameter and 20 mm high and inserted into the W&P cell. The cap was set in place, a vertical load was applied, and the specimen compressed to its final height. Thereafter, the cap was tightened to the cell, the tensiometer was installed through the cap, and finally the pressure chamber was assembled and connected to the air supply system.

The air pressure, both in the tank and in the pressure chamber, was kept at atmospheric value for four days. This period of time was sufficient for the matrix suction to attain a constant value, and for the net stress to reduce significantly its rate of decay with time. Thereafter, three different values of air pressure increments (200, 400 and 600 kPa) were applied and maintained constant for one day each. After the one-day period, the air pressure was reduced back to atmospheric and again kept constant for another day, as shown in Figure 3.

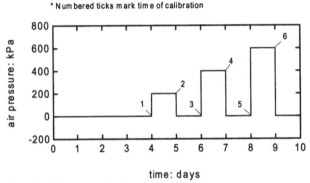

Figure 3. Air pressure loading sequence.

While performing these preliminary tests, an interaction (arching effect) between the soil specimen and the measurement diaphragm of the net stress transducers was observed. As a result, the changes in net stress recorded by the flush diaphragm transducers were underestimated. To correct these measures, a series of calibrations were performed by keeping constant the air pressure inside the sample and, at the same time, modifying the air pressure outside the cell in small increments. Six calibration tests were performed before any increase or reduction of air pressure in the tank and the pressure chamber. Details of the calibration procedure are given in Tarantino et al. (2000a).

5.2 *Oedometer tests*

The bronze porous disc was first saturated in boiling water and assembled into the oedometer base underwater to prevent desaturation. Still underwater, the two semi-permeable membrane sheets were put on the porous disc, one on top of the other. After clamping the semi-permeable membranes, the oedometer base was removed from the water basin where it was assembled and put in place in the loading frame.

The glass bottle, already connected to the oedometer base through the flexible tubes of the peristaltic pump, was positioned 300 mm below the oedometer. At this stage, the bottle was filled only with water. Since water pressure in the oedometer base was less than atmospheric, possible breaks in the semi-permeable membrane could be detected by ingress of air into the system.

The soil specimen was then trimmed using the cutting edge oedometer ring and set in place onto the base. The top cap was positioned on the sample and a vertical pressure of 50 kPa ap-

8

plied to ensure contact between the soil sample and the top cap. The two tensiometers were installed through the cap after applying a soil paste on the tensiometer porous stone.

The bottle containing water was replaced with another bottle containing the previously prepared osmotic solution. Penicillin was added to the solution to prevent bacterial growth on the semi-permeable membrane. The concentration of PEG was selected such as to generate a matrix suction slightly less than the initial suction of the sample. As a result, the specimen initially tended to expand to conform to oedometer conditions.

The pump was switched on and the solution circulated underneath the specimen to impose the given negative pore water pressure. All tests were initiated only after equilibrium was attained. Tests were run by increasing or decreasing the total vertical stress in steps and allowing the specimen to reach equilibrium under each applied stress. Equilibrium conditions were checked in terms of vertical strain, pore-water pressure, and water content.

Specimens were loaded in steps up to a total stress of 1600 kPa and then unloaded. The maximum increment in total stress was such as to never exceed the value of the negative pore water pressure. This was intended to avoid generation of positive pore pressures and subsequent extrusion of soil through the annular gap between the cap and the inner oedometer ring. The maximum decrement in total stress was such as to avoid lowering the pore water pressure to values less than −1200 kPa. Beyond this value, there would be a potential risk of cavitation in the two tensiometers used.

6 EXPERIMENTAL INVESTIGATIONS IN THE RANGE OF CONTINUOUS AIR PHASE USING THE W&P APPARATUS

A series of six tests was performed in the range of continuous air phase using the W&P apparatus. The results of these tests are reported in detail in Tarantino et al. (2000a). In the work presented herein, some experimental data are emphasised in order to compare these tests to those carried out at a high degree of saturation.

The degrees of saturation and the corresponding matrix suctions investigated at this stage are reported in Figure 4, and compared to those explored by Bishop & Donald (1961) in silt, and by Fredlund & Morgenstern (1977) in kaolin. Values of degree of saturation and matrix suction of the six specimens examined are consistent, in the sense that degree of saturation decreases with increasing suction. This is not the case for the specimens tested by Fredlund & Morgenstern (1977). However, as pointed out by the authors before, during some tests leakage of water occurred through the rubber membrane surrounding the sample, and this probably explains these discrepancies.

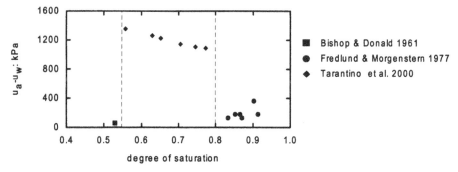

Figure 4. Degrees of saturation and matrix suctions investigated in this programme.

Typical results of a null test are presented in Figure 5, where air pressure, matrix suction, and net stresses, measured on the top and bottom of the specimen, are plotted versus time. Data are presented with reference to the specimen having the highest degree of saturation (S = 0.77).

This will be helpful in interpreting the results from the specimen with 0.88 degree of saturation, which are presented in next section.

Measures of lateral stresses are not reported in Figure 5. In fact, these measures were not accurate and, hence, not used to interpret test results. Nevertheless, they were useful to check that no lateral contraction occurred during testing.

The large variations recorded immediately after increasing or decreasing the air pressure are due to the particular procedure adopted to modify air pressure, both in the tank and in the chamber. A change in air pressure in the chamber is recorded almost instantaneously by the transducers. On the contrary, a pressure change in the tank is recorded with a time lag due to equalisation of air pressure inside the sample.

If the pressure in the tank and in the chamber were both modified instantaneously to the desired final value (200, 400 or 600 kPa), the tensiometer would be initially subjected to an unbalanced pressure that could eventually damage it. To prevent this, the two air pressures were modified in steps of 50 kPa and in a non-simultaneous fashion, according to the sequence shown in Figure 6. Valve G (Fig. 1) was repeatedly open and closed in order to limit the changes in pressure recorded by the tensiometer in a range of about 50 kPa.

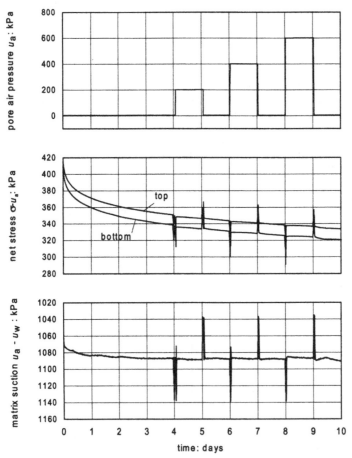

Figure 5. Null test on the specimen having a degree of saturation of 0.77.

Only when pressure in the tank and the chamber reached their final value, and pore air pressure attained equalisation, matrix suction was actually measured by the tensiometer.

Test data shown in Figure 6 made it possible to verify that air phase was continuous in all six specimens tested. It may be noticed that complete air pressure equalisation was attained in about 1 h for the specimen having a degree of saturation of 0.77. The same results were obtained for the specimen with the lowest degree of saturation (S = 0.56), for which air was assumed to be continuous in the voids. As a result, air was continuous in the entire range of degree of saturation investigated in this work (0.56 - 0.77).

As shown in Figure 5, matrix suction remained almost unaffected by the changes in pore air pressure, with similar results found in all tests. The recorded changes in matrix suction, expressed as a percent of the value measured at atmospheric pore air pressure, are reported in Figure 7. These changes were always less than 0.5 % and slightly greater than measurement accuracy (0.1 %).

Measurements of net stress did not remain constant during the test and decayed with time. As such, the main problem in the interpretation of these data was to distinguish the changes caused by pore air pressure variations from those due to stress relaxation. To this end, the relaxation curve corresponding to atmospheric air pressure was first estimated by interpolating net stress data recorded at atmospheric air pressure. Secondly, changes in net stress induced by changes in pore air pressure were calculated using this curve as datum (Fig. 8a). Thereafter, measures were corrected to account for errors due to interaction between soil and net stress transducers' dia-

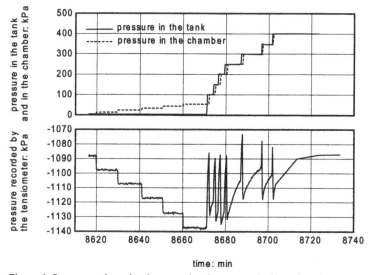

Figure 6. Sequence adopted to increase the air pressure in the tank and chamber to 400 kPa (data relative to the specimen with 0.77 degree of saturation).

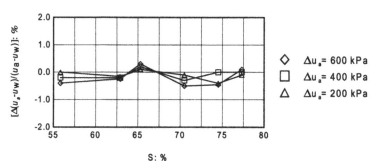

Figure 7. Matrix suction changes.

phragms (Fig. 8b). The complete set of results is shown in Figure 9. Net stress changes were rather unaffected by the degree of saturation and appear to increase with the magnitude of the air pressure variation. The changes in net stress, expressed as a percent of the value measured at the end of the test, were always less than 2%, with the exception of one data (equal to 2.4%). Although these changes are somewhat greater than the measurement accuracy, which is better than 0.13%, they are still deemed negligible for most practical purposes.

Based on the above results, it seems reasonable to assume that the three stress variables σ, u_w, and u_a can be combined in two effective stress fields for the case where the air phase is continuous in the voids.

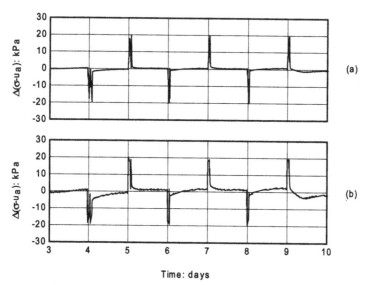

Figure 8. Net stress changes at the bottom: (a) measured; (b) after correction.

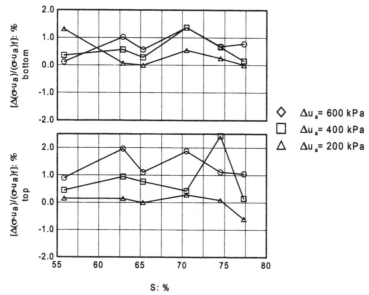

Figure 9. Net stress changes.

12

7 EXPERIMENTAL INVESTIGATIONS IN THE RANGE OF DISCONTINUOUS AIR PHASE USING THE W&P APPARATUS

A *null* test was carried out on a sample having a degree of saturation of 0.88 after compression in the W&P cell. As shown in Figure 10, the test lasted about 20 days and only air pressure increments of 200 and 400 kPa were applied. Each loading step lasted much longer because air was occluded within the pore water and, hence, pore air pressure equalised much slower. This is clearly emphasised by the response of the tensiometer while increasing air pressure from 0 to 400 kPa (Fig. 11). It may be observed that air pressure inside the specimen equalised in a period of about 1000 min. This time is about 17 times greater than that required for equalisation in the case of continuous air phase. This result may be taken as the evidence that air phase was indeed occluded in the pore water.

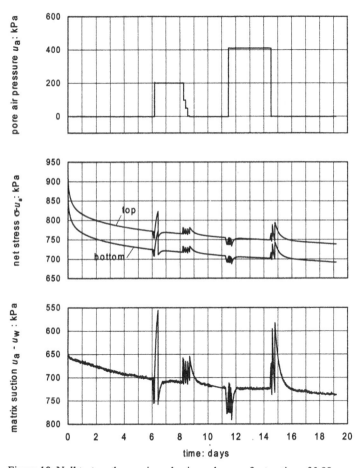

Figure 10. Null test on the specimen having a degree of saturation of 0.88.

In Figure 10, it can be noticed that matrix suction did not attain a constant value during the 20-day test, although it almost equalised at the end of the test. This slow equalisation was probably due to evaporation of soil water into the tank. The rate of decay decreases with time since the evaporated soil water tends to saturate the water vapour in the tank.

13

Such evaporation probably occurs also in the case of continuous air phase. However, in this case, all pores in the specimen are exposed to air, and the water is removed uniformly from the soil. This causes a negligible drop in suction that remains almost constant during the test. On the contrary, when air is discontinuous, evaporation is concentrated in the proximity of the porous discs. The curvature of water menisci significantly reduces and, hence, matrix suction increases.

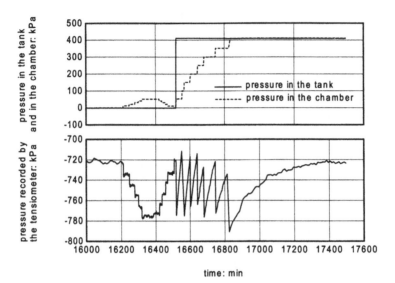

Figure 11. Sequence adopted to increase the air pressure in the tank and in the chamber to 400 kPa (data relative to the specimen with 0.88 degree of saturation).

Since matrix suction decayed with time, changes in matrix suction due to air pressure variations were determined according to the same procedure followed to interpret net stress data. These changes are not greater than 0.5 % of the measured matrix suction (Fig. 12), similarly to the case of continuous air phase.

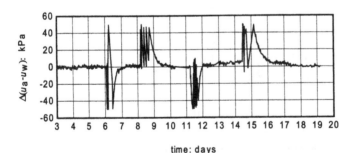

Figure 12. Matrix suction changes for the test at high degree of saturation.

Net stress changes were determined following the same procedure, and results are shown in Figure 13. Yet, no differences were found with respect to the case of continuous air phase. In particular, changes in net stress, expressed as a percent of the value measured at the end of the test, are less than 1%. This result suggests that the three stress variables σ, u_w, and u_a can be

14

combined in two effective stresses even when the air phase is discontinuous. Accordingly, axis translation technique could be also used when air phase is occluded, with the only limitation of a much slower equalisation period for the air pressure.

Although the experimental procedure has been proven to be reliable, since it has been validated with respect to the known case of saturated soils (Tarantino et al. 2000a), results from one test are not sufficient to draw any conclusions and further testing is required.

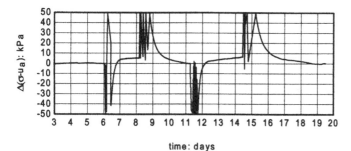

time: days

Figure 13. Net stress changes at the top for the test at high degree of saturation (after correction).

8 OEDOMETER TESTING AT ATMOSPHERIC AIR PRESSURE

Two oedometer tests were conducted imposing matrix suctions of 250 and 850 kPa. Degrees of saturation remained approximately constant during testing, attaining values in the range of 0.94-0.95 and 0.91-0.92, respectively.

Results from the test on the specimen having S = 0.95 are presented in Figure 14, where total vertical stress, temperature, matrix suction, total and water volume changes are plotted versus time.

It is worth noticing the good response of the two tensiometers. The two measurements are in good agreement and differences remain within a range of 10 kPa. Repeatability better than ± 5 kPa is typical for the IC tensiometer. This has been observed in many other measurements where two or more tensiometers were simultaneously used on the same sample.

It is also important to note the significant fluctuations recorded by the tensiometer *pr4* when laboratory air temperature changed significantly due to malfunctioning in the air conditioning system. These fluctuations are in phase with temperature changes, and this type of response was generally recorded when the tensiometer porous stone was not fully saturated. On the contrary, the other tensiometer was insensitive to temperature changes, and this was attributed to a good saturation of its porous stone. A good saturation could be achieved by applying repeated cycles of cavitation and subsequent pre-pressurisation (Tarantino et al. 2000b).

It is also worth noticing that the tensiometers remained in contact with the top surface of the specimen during both loading and unloading. The system devised to install the tensiometer through the top cap, has then proved to be satisfactory.

Water losses due to evaporation were not eliminated despite the many precautions taken. The osmotic solution was placed in a closed bottle instead of a beaker covered with silicon oil as suggested by Dineen & Burland (1995). The annular gap between the cap and the inner oedometer ring was covered with an elastomer membrane obtained by cutting and then assembling different parts of a Bellofram membrane.

In Figure 15, the difference between total and water volume change is plotted versus time. It can be noticed that this difference increases almost linearly with time, with either the specimen being loaded or unloaded. This may be attributed to water evaporative losses through the flexible tubes of the peristaltic pump and the junctions of the elastomer membrane over the annular gap. In a first approximation, evaporation rate was assumed to be constant throughout the test and estimated on the basis of the data shown in Figure 15. Water volume changes were then recalculated to account for evaporation (Fig. 14).

15

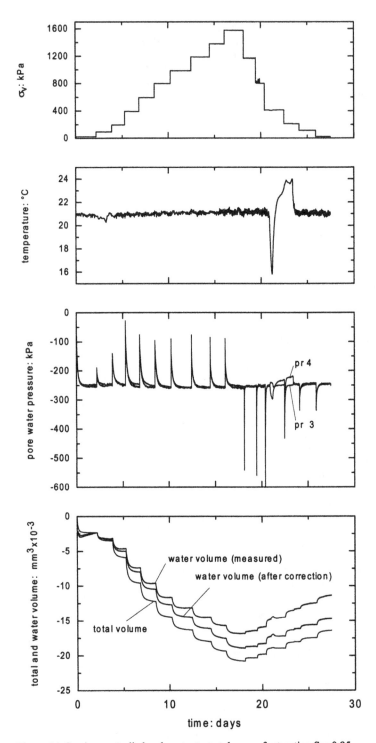

Figure 14. Suction controlled oedometer test at degree of saturation S = 0.95.

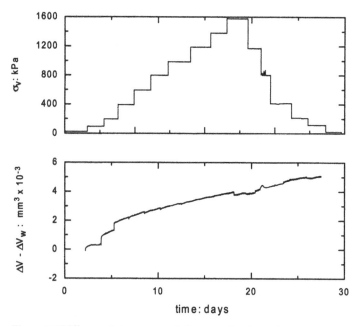

Figure 15. Difference between recorded water and volume changes.

The results of the two oedometer tests at constant suction are reported in Figure 16, where the void ratio e (void volume to solid volume ratio) and the water ratio e_w (water volume to solid volume ratio) are plotted against the effective stress for saturated soils σ-u_w. Compression curves should not differ if the principle of effective stresses for saturated soils still remains applicable for unsaturated soils at high degrees of saturation.

As it would be expected, the oedometer curves in terms of e and e_w are not coincident since the degree of saturation is less than unity (S= e_w/e). Curves in terms of e tend to remain superposed, while the curves representing e_w progressively detach from the e curve as matrix suction increases, i.e. degree of saturation decreases. All curves seem to be almost parallel in a semilogarithmic scale, at least in their virgin region.

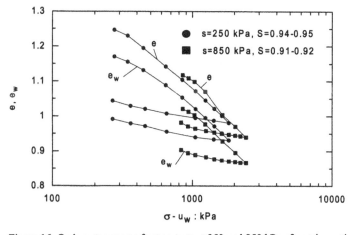

Figure 16. Oedometer curves for two tests at 250 and 850 kPa of matrix suction.

However, no conclusions can be drawn at this stage of the study. The authors take these results very cautiously since uncertainties still exist about water content changes and estimates of initial void ratios. Besides, these curves are somehow difficult to compare because the loading and unloading steps were different. Increment in total vertical stress had not to exceed the applied suction in order to avoid generation of positive water pressures. This would have caused soil extrusion through the cap and the inner ring. Strain controlled tests are indeed more appropriate than stress controlled tests. Rate of strain can be maintained slow enough to ensure a constant suction throughout the test. Unfortunately, a press was not available at the time of the tests, which were therefore carried out controlling vertical stress.

These preliminary results show, however, that the osmotic controlled oedometer is suitable to carry out the proposed investigation. Matrix suction was well controlled using the osmotic technique and accurately measured with the tensiometers. The balance allowed continuos monitoring of water content changes and the insertion of the bronze porous disc made it possible an accurate measurement of vertical settlements.

9 CONCLUSIONS

The majority of experimental data on unsaturated soils have been obtained using axis translation technique. The validity of this technique lies on the assumption that the total stress, the pore-water pressure, and the pore-air pressure can be reduced to two effective stresses. This assumption is however supported by limited experimental evidence. Although it has been generally assumed that axis translation technique is applicable in the case of continuos air phase, just one case study has been presented in the literature for this particular unsaturated state. Besides, no data are available in the negative range of pore water pressure, despite its relevance in engineering practice

An experimental programme has been underway to investigate the effective stresses for unsaturated soils in the negative range of pore water pressure. For this purpose a new apparatus has been designed and implemented.

A first set of tests was carried out to investigate the range of continuous air phase. Experimental results confirm that two effective stresses, for example net stress and matrix suction, control the behaviour of unsaturated soils in this range of degrees of saturation.

Using the same apparatus, one test was then carried out on a specimen at high degree of saturation, where air phase was occluded. Results would suggest that axis translation technique remains applicable even at high degrees of saturation, the main limitation being the long time required for equalisation. However, at this stage no conclusions can be drawn and more tests are required.

To investigate whether the principle of effective stress for saturated soils can be extended to unsaturated soils at high degrees of saturation and, hence, whether the three stress variables can be reduced to a single one, an osmotic controlled oedometer was set up. This apparatus has proved suitable to study the transition from saturated to unsaturated state without increasing air pressure. Two IC tensiometers were used to measure soil matrix suction and their response was more than satisfactory. The design of the oedometer base made it possible to measure vertical settlements accurately. Other experimental problems, however, still remain to be solved. In particular, soil water evaporation needs to be controlled and initial void ratio more accurately determined. Besides, strain controlled tests have to be conducted to ensure a constant suction throughout the test.

ACKNOWLEDGEMENTS

The authors wish to thank Dr. Laureano R. Hoyos Jr of the University of Texas at Arlington, USA, for reviewing the paper.

REFERENCES

Alonso, E.E., A. Gens & D.W. Hight 1987. Special problems soils. General report. *Proc. 9th Eur. Conf. Soil Mech. Geotech. Eng., Dublin,* 3: 1087-1146.

Bishop, A.W. 1959. The principle of effective stress. *Teknisk Ukeblad* 106(39): 859-863.

Bishop, A.W. & I.B. Donald 1961. The experimental study of partly saturated soil in the triaxial apparatus. *Proc. 5th Int. Conf. Soil Mech., Paris* 1:13-21.

Bocking, K.A. & D.G Fredlund 1980. Limitations of the axis translation technique. *Proc. 4th International Conference on Expansive* Soils, *Denver:* 117-135.

Bolzon, G. & B.A. Schrefler 1995. State surfaces of partially saturated soils: an effective pressure approach. *Appl. Mech. Rev.* 48(10): 643-649.

Cui, Y.J. & P. Delage 1996. Yielding and plastic behaviour of an unsaturated silt. *Géotechnique* 46 (2): 291-311.

Delage, P., G.P.R. Suraji De Silva & E. De Laure 1987. Un nouvel appareil triaxial pour les sols non-saturés. *Proc. 9th ECSMFE, Dublin* 1: 25-28.

Dineen, K. & J.B. Burland 1995. A new approach to osmotically controlled oedometer testing. In E.E. Alonso & P. Delage (eds.), *Unsaturated Soils, Proc. 1st Int. Conf. on Unsaturated Soils, Paris* 2: 459-465. Rotterdam: Balkema.

Fredlund, D.G. & N.R. Morgenstern 1977. Stress state variables for unsaturated soils. *J. Geotech. Engng. Div. Am. Soc. Civ. Engrs.* 103(GT5): 447-466.

Geiser, F., L. Laloui & L. Vuilliet 2000. Modelling the behaviour of unsaturated silt In *Experimental Evidence and Theoretical Approaches in Unsaturated Soils; Proc. of an International Workshop, Trento, 10-12 April 2000, this issue.*

Geiser, F. 2000. Applicability of a general effective stress concept to unsaturated soils. *Proceedings of the Asian Conference on Unsaturated Soils, 18-19 May 2000, Singapore:* 101-105. Rotterdam: Balkema.

Kahlili, N. 2000. Application of the effective stress principle to volume change in unsaturated soils. *Proceedings of the Asian Conference on Unsaturated Soils, 18-19 May 2000, Singapore:* 119-124. Rotterdam: Balkema.

Khalili, N. & M.H. Kabbaz 1998. A unique relationship for χ for the determination of the shear strength of unsaturated soils. *Géotechnique* 48: 681-688.

Jommi, C. 2000. Remarks on the constitutive modelling of unsaturated soils. In *Experimental Evidence and Theoretical Approaches in Unsaturated Soils; Proc. of an International Workshop, Trento, 10-12 April 2000, this issue.*

Mongiovì L. & A. Tarantino 1998. An apparatus to investigate on the two effective stresses in unsaturated soils. *Proc. 2nd Int. Conf. on Unsaturated Soils, Beijing, China:* 422-425. Beijing: International Academic Publishers.

Pietruszczak, S. & G.N. Pande 1996. Constitutive relations for partial saturated soils containing gas inclusions. *Journal of Geotechnical Engineering* 122(1): 50-59.

Ridley, A.M. & J.B. Burland 1993. A new instrument for the measurement of soil moisture suction. *Géotechnique* 43(2): 321-324.

Ridley, A.M. & J.B. Burland 1996. A pore pressure probe for the *in situ* measurement of soil suction. In Craig (ed.), *Advances in site investigation practice:* 510-520. London: Thomas Telford.

Tarantino, A. 1998. *Le variabili di stato tensionale per i terreni non saturi.* PhD Thesis, Politecnico di Torino, Italy.

Tarantino, A. & L. Mongiovì 1999. Impiego della tecnica osmotica per il controllo della pressione negativa dell'acqua interstiziale. In *Atti del XX Convegno Italiano di Geotecnica, Parma, Italy:* 295-300. Bologna: Pàtron Editore.

Tarantino, A. & L. Mongiovì 2000. A study of the efficiency of semipermeable membranes in controlling soil matrix suction using the osmotic technique. *Proceedings of the Asian Conference on Unsaturated Soils, 18-19 May, Singapore:*303-308. Rotterdam: Balkema.

Tarantino, A., L. Mongiovì & G. Bosco 2000a. An experimental investigation on the isotropic stress variables for unsaturated soils. *Géotechnique* 50(3): 275-282.

Tarantino, A., G. Bosco & L. Mongiovì 2000b. Response of the IC tensiometer with respect to cavitation. *Proceedings of the Asian Conference on Unsaturated Soils, 18-19 May 2000, Singapore:* 309-314. Rotterdam: Balkema.

Terzaghi, K. 1936. The shearing resistance of saturated soils and the angle between the planes of shear. *Proc. 1st Int. Conf. Soil Mech. Found. Eng., Camb., Mass.* 1: 54-56.

Experimental Evidence and Theoretical Approaches in Unsaturated Soils, Tarantino & Mancuso (eds)
© *2000 Taylor & Francis, ISBN 90 5809 186 4*

Investigating infiltration into compacted unsaturated heterogeneous soils

J. Shevelan & C. Hird
Department of Civil and Structural Engineering, University of Sheffield, UK

ABSTRACT: Many simulations of infiltration are based on the assumption of homogeneous soil water characteristics, which in many cases will not be accurate. In the case of a compacted natural clay landfill liner or cap, variations are inevitable and may have a significant effect on its performance. Using a combination of numerical and laboratory work the influence of spatial variation of soil properties on infiltration has been investigated. A finite difference model, using the Richards equation, has been used to simulate infiltration into layered soils. This has been compared to laboratory infiltration experiments into layered samples. A two-step stochastic methodology has also been developed to assess the influence of variations in initial moisture content and permeability on infiltration.

1 INTRODUCTION

In the last decade there has been an increasing focus on the risk of contamination of groundwater. One of the more clearly identifiable processes where there is a risk of pollution is waste disposal to landfill sites. To provide guidance on minimising the risk of contamination from landfill sites, attempts have been made to understand and model the physical processes involved in the infiltration of leachate through the basal liner into the surrounding environment (e.g. Benson and Daniel, 1994). Previous work has mainly relied on the assumption of the liner being saturated and homogeneous. However, these assumptions may not be realistic. A compacted clay liner will not, initially, be saturated and, no matter how carefully constructed, will contain spatial variations of soil properties. Sometimes these will occur in layers through a variation of material or compaction effort in successive lifts. It is therefore desirable to model the effects of unsaturated flow and soil heterogeneity on infiltration into landfill liners. Such an approach is even more justified for predicting infiltration into landfill capping systems.

The overall aim of the research described in this paper was to investigate the influence of spatial variation of soil properties on infiltration through unsaturated soils in the context of landfill liners and caps. The research employed a combination of numerical and laboratory work which is outlined below and described in detail by Shevelan (2000).

2 EXPERIMENTAL METHODS

The experimental work was primarily designed to provide infiltration test data against which the numerical work could be validated. To eliminate unwanted variations, rather than use natural soils, two artificial soils were created and compacted at their optimum moisture contents. The two soils comprised Leighton Buzzard Grade D sand (300-150 μm) mixed with 10% kaolin (material C) and 30% kaolin (material F). For each material a compaction test was carried out and the bulk properties derived, as given in Table 1. Throughout the experimental work, the tabulated values were compared with measurements on test specimens to check that the specimens were compacted correctly and to ensure that any variations in the material properties were acceptable.

Table 1. Bulk properties of materials F and C

Material	Optimum gravimetric moisture content	Maximum dry density (Mg/m3)	Minimum void ratio	Minimum porosity	Volumetric moisture content, θ	Degree of saturation
F	11%	1.91	0.39	0.28	0.21	0.75
C	10%	1.73	0.53	0.35	0.17	0.50

Triaxial permeability tests were carried out on compacted specimens (70 mm high and 100 mm in diameter) to determine the saturated permeabilities of the two materials.

The retention characteristics of the materials were determined using a multi-step pressure plate method, employing the axis-translation technique developed by Hilf (1956). In this multi-step method (Fourie and Papageorgiou, 1995) the air pressure is applied to the specimen in increments but for each increment, instead of allowing the specimen to reach hours. The specimen then reaches equilibrium rapidly and the suction can be determined by measuring the water pressure beneath the pressure plate. This method allows many points on the retention curve to be determined from one specimen in a relatively short time. The van Genuchten (1980) equation was fitted to the experimental data using the RETC computer program (van Genuchten et al., 1991).

As drainage occurred under each air pressure increment in the pressure plate tests, the unsaturated permeability was also determined using the method developed by Gardner (1956), Miller and Elrick (1958) and Kunze and Kirkham (1962). This method, also employed more recently by Fourie and Papageorgiou (1995), relies on graphical fitting of the outflow rates to theory. The results were compared to unsaturated permeabilities calculated using the Mualem (1976) function together with the van Genuchten fitting parameters for the retention curve.

The infiltration tests were carried out in series of cylindrical Perspex cells (100mm in diameter and 100mm high), Figure 1. Specimens were compacted into the individual cells and then these were bolted together. This allowed both homogenous (e.g. CC or FF) and heterogeneous soil columns to be formed, depending on the material placed in each 100mm layer (e.g. CFC or FCF). The layering was reflected in the test reference number, e.g. CFC in Figure 6. A Mariotte system, with a burette arrangement to determine the infiltration rate, was used to supply a constant head of 0.64m to the top of the column. The burette was also used to carry out a falling head permeability test after infiltration of the homogeneous columns. The bottom of the soil column was open to atmosphere to allow free drainage. Subsequently, it was realised that it would have been better to apply a constant suction in order to control the experiment more precisely and to facilitate the numerical modelling. Tensiometers were used to measure the initial soil suction and to observe the advance of the wetting front through the column. The air pressure in one layer was also monitored during infiltration using a sensitive pressure transducer. The final degree of saturation of the column was determined by dissection.

3 NUMERICAL METHODS

The numerical work was based a version of the widely accepted Richards (1931) equation, describing one-dimensional unsaturated flow.

$$\frac{\partial \theta}{\partial t} = \frac{\partial}{dz}\left(k(\theta)\frac{\partial h}{\partial z}\right) - \frac{\partial k}{\partial z} \tag{1}$$

where h = pore water pressure head, z = depth, k(θ) = unsaturated permeability, θ = volumetric moisture content and t = time.

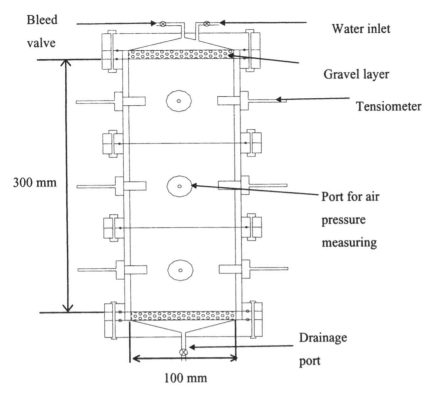

Figure 1. Schematic diagram of infiltration test rig

An implicit finite difference formulation of the Richards equation was used to form a series of tridiagonal equations for the soil column, subdivided in height. The series of equations, written in terms of nodal heads (h), was then solved to give heads at the next time step by direct elimination using the Thomas algorithm (Remson et al. 1971) and thus the solution was marched forward in time. The numerical solutions were produced by writing a computer program in FORTRAN which included routines to compute the current values of θ and k(θ) from h using appropriate mathematical functions. To monitor the accuracy of the solutions, mass balance error (MBE) calculations were included. The model was initially validated against the numerical solutions and experimental data for one-dimensional infiltration in homogeneous soil presented by Haverkamp et al. (1977). Further validation was sought by applying it to the laboratory infiltration tests. The van Genuchten (1980) retention curve fitted to the experimental data (Figures 2 and 3) and the Mualem (1976) permeability function were employed in these simulations.

4 STOCHASTIC MODELLING

A stochastic modelling methodology was developed with the intention of investigating how variation in the physical properties of compacted materials placed in a landfill liner or cap affects infiltration. This development was based on information, firstly, that samples taken from actual landfill liners exhibit log normal distributions of saturated permeability (Rogowski, 1990, Donald and McBean, 1994) and, secondly, that there is a systematic relationship between gravimetric moisture content at compaction and saturated permeability.

The two step method developed by Loll and Moldrup (1998) was used to investigate how the initial gravimetric moisture content of a compacted soil, and hence its permeability, affect infiltration. In the first step of the method, deterministic response surfaces (DRSs) are created depicting the volumetric moisture contents at selected depths and times for all feasible combinations of gravimetric moisture content and saturated permeability. At present the retention characteristics are assumed to be independent of the initial gravimetric moisture content, but the method could be refined in this respect. The second step involves interpolation on the DRSs to obtain the responses for each pair of input parameters selected using a Monte Carlo sampling scheme (applied to specified distributions of gravimetric moisture content and saturated permeability). This method is considerably more efficient than a standard Monte Carlo simulation; the number of infiltration simulations was reduced from about 10000 to just 100. The results generated were then used to predict a statistical distribution of the volumetric moisture content at the chosen depths and times.

The methodology was tested using input distributions, and a relationship between gravimetric moisture content and saturated permeability, derived from some actual landfill site data. However, because corresponding retention data were unavailable, these were used in combination with retention data obtained for material C. An unrealistically high average saturated permeability was also used to speed up the computations, since there were limits to the size of the time steps that produced stable solutions. Therefore, at this stage the methodology has not been applied to a real landfill situation.

5 RESULTS AND DISCUSSION

In the triaxial permeability tests the permeability was observed to vary with the effective stress level. Extrapolating the trend to very low stress levels, as existed in the infiltration tests, gave values of saturated permeability of $7.0*10^{-9}$ m/s and $5.0*10^{-6}$ m/s for materials F and C respectively.

Desorption curves for materials F and C, obtained using the multi-step pressure plate method, are shown in Figures 2 and 3. The fitted van Genuchten retention curves are also shown and the parameter values are given in Table 2.

Table 2. Retention curve fitting parameters for materials F and C

Material	θs	θr	$\alpha(m-1)$	a	b
F	0.28	0.14	2.115	1.234	0.1898
C	0.35	0.09	1.978	2.176	0.5404

The unsaturated permeability results for material C, also obtained from the pressure plate tests, are shown in Figures 4 and 5. It should be noted that, to save on preparation effort, the test specimens used for the triaxial permeability tests were reused in these pressure plate tests. Because the specimens had already been subjected to significant effective stresses (15-20 kPa), their initial saturated permeability is likely to have been lower than the value quoted above, probably about $2.7*10^{-6}$ m/s. This value is marked on Figures 4 and 5 to give an indication of the change in permeability with loss of saturation or increase of suction. For material F, where freshly prepared specimens were used, the experimental data were more scattered but showed a similar trend. In Figure 5 there is a marked difference between experimental and predicted permeabilities, with the Mualem (1976) permeability function underestimating the unsaturated permeabilities significantly. This is in line with evidence provided by Chiu and Shackleford (1998). There was a similar mismatch between the experimental data and the predictions for material F

All the infiltration tests were continued until a steady state appeared to have been reached. Upon dissection the soil columns showed a recognisable upper saturated zone 1-2 cm deep with a sharp transition to a less saturated lower region, the transmission zone. The degree of saturation of the transmission zones showed reasonably good consistency from test to test with material C reaching 70-80% saturation and material F reaching 85-95%. This is illustrated in Figure 6. The permeabilities of the materials derived from the falling head tests at the end of infiltration were $5.7*10^{-7}$ m/s and $3.3*10^{-9}$ m/s for materials C and F respectively.

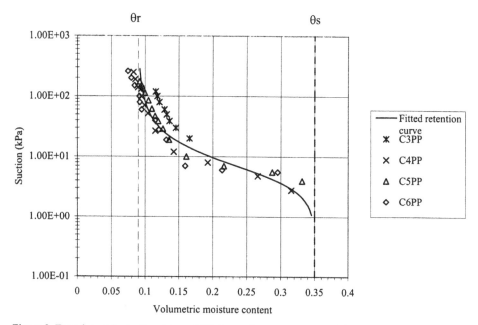

Figure 2. Experimental retention data and fitted retention curve compared for material C

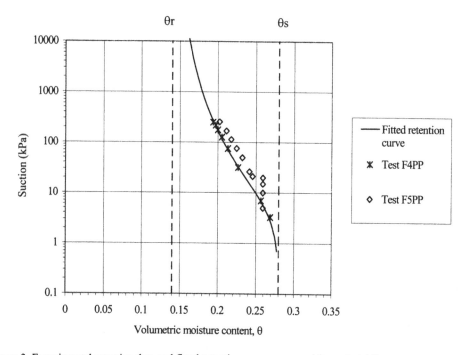

Figure 3. Experimental retention data and fitted retention curve compared for material F

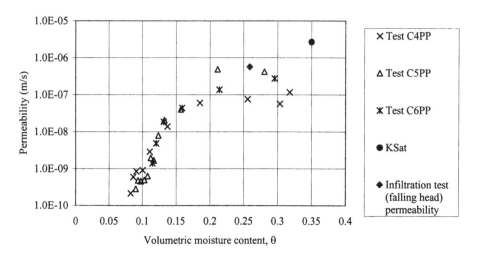

Figure 4. Unsaturated permeability against volumetric moisture content for material C

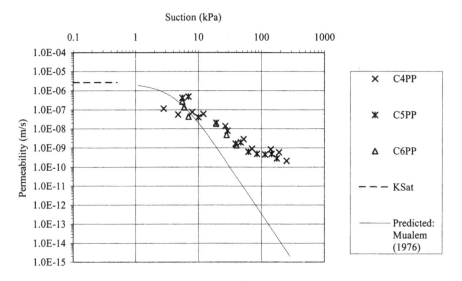

Figure 5. Unsaturated permeability against suction for material C

Assuming that the average final degree of saturation for material C was 75% (θ=0.26), in Figure 4 it can be seen that the former value shows reasonable agreement with the unsaturated permeabilities obtained from the pressure plate tests. This was also true for material F. While conducting the infiltration tests it was recognised that preferential flow paths, or "fingering" of the wetting front, might develop. The final gravimetric moisture content distributions showed no evidence of this, although it is possible that fingering developed at an intermediate stage of the tests. There was also no evidence of air compression ahead of the wetting front.

Figures 7 and 8 show both simulated and experimental results for infiltration into homogenous two-layer soil columns (CC and FF). Spurious spikes occasionally occurred in the experimental data due to voltage fluctuations and these should be ignored. Simulations were carried out using saturated permeabilities set equal to the permeability values obtained from both the triaxial test and the falling head test conducted at the end of infiltration. Using the falling head permeability incorporates, rather crudely, the effects of reduced saturation found experimentally. The CC simulations gave better agreement with the experimental data when the falling head permeability was used, while the FF simulations gave marginally better agreement when the triaxial permeability was used. However, in the latter case the difference in permeabilities had relatively little effect. Overall neither the CC nor the FF simulation produced good agreement with the experimental results. It is likely that the differences between experiment and simulation are primarily the result of using the desorption retention curve to model a sorption process. In addition, unsaturated permeabilities may have been poorly predicted by the Mualem (1976) function and there were difficulties in modelling the lower boundary condition adequately, particularly in the FF simulation.

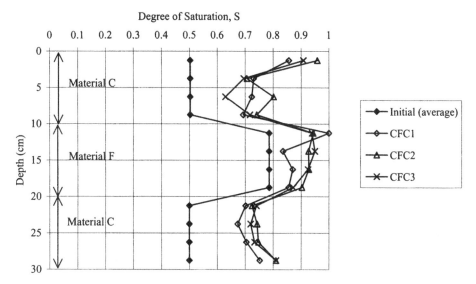

Figure 6. Initial and final degree of saturation profiles for CFC infiltration tests

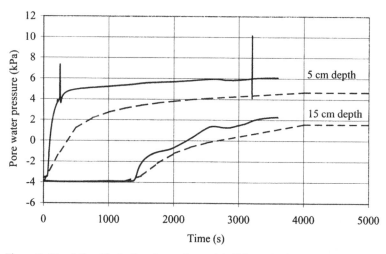

Figure 7. Simulation (dashed) and experimental (solid) results for CC infiltration test using $K_{sat}=5.7*10^{-7}$ m/s

For a heterogeneous soil column, the three-layer CFC simulation shown in Figure 9 agreed reasonably well with experimental results. However, a simulation of the FCF case showed very poor correlation with the experimental data.

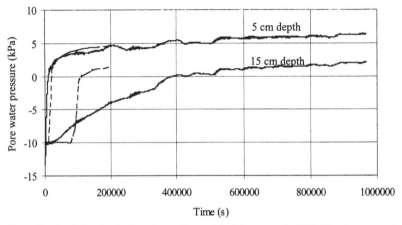

Figure 8. Simulation (dashed) and experimental (solid) results for FF infiltration test using $K_{sat} = 7.0*10^{-9}$ m/s

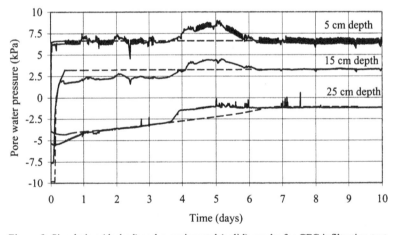

Figure 9. Simulation (dashed) and experimental (solid) results for CFC infiltration test

The stochastic model was run for infiltration into a 300mm thick soil layer subjected to the same boundary conditions as used in simulating the infiltration tests. An example of a DRS from the first step of the method is shown in Figure 10 where the initial relative permeability refers to the ratio of the saturated permeability to an arbitrary datum value taken as the saturated permeability of material C. The Monte Carlo sample used in the second step is depicted in Figure 11 and the interdependence of the two input variables is evident. The DRSs were used to generate model responses for each element of this sample, i.e. for each pair of input parameters. Analysis of the stochastic outputs was carried out by creating cumulative distribution functions (CDFs) of volumetric moisture content at different times and depths. The CDFs were then used to obtain volumetric moisture contents corresponding to the 2.5%, 50 % and 97.5% percentiles, as shown for one depth in Figure 12.

Results such as those shown in Figure 12 can be used to help predict the rate of infiltration with given distributions of input parameters. Such predictions could form part of a risk based analysis for landfill simulation.

In much of the numerical modelling, including the comparisons made with Haverkamp et al. (1977) which are not reported here, the MBEs were within acceptable limits. For example, for the CC case MBEs were less than 1%. However, higher and sometimes erratic errors were present in the modelling of the FF, CFC and FCF cases. At times errors of 20-30% occurred and this must put the accuracy of the modelling in doubt. Further work is needed to establish the precise cause of these errors and to develop a superior solution technique.

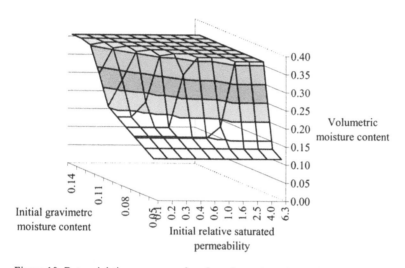

Figure 10. Deterministic response surface for volumetric moisture content at 15cm depth after 1800 seconds

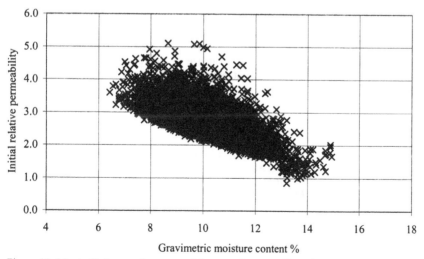

Figure 11. Monte Carlo sample generated from random sampling of a normal moisture content distribution and a log normal permeability distribution

29

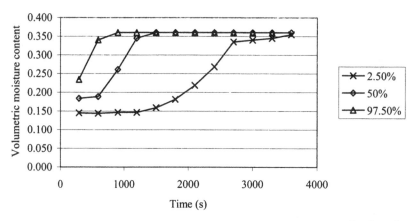

Figure 12. Stochastic model response showing 2.5%, 50% & 97.5% percentiles for volumetric moisture content against time at a depth of 15 cm

6 CONCLUSIONS

This paper has described an attempt to develop and validate a methodology for modelling infiltration into a compacted clay liner or cap and the effect of spatial variations of soil properties. The following main conclusions can be drawn.

(1) A numerical model based on the Richards equation was successfully validated against results in the literature (Haverkamp et al., 1977). However, it did not perform especially well in simulating new infiltration experiments conducted on compacted sand-kaolin mixtures. A major reason for this was probably the use of a desorption retention curve instead of a sorption curve in the model.

(2) The desorption retention characteristics and unsaturated permeabilities of the materials used in the infiltration tests were successfully determined using the pressure plate apparatus and the methods described by Fourie and Papageoriou (1995). As suction increased the measured permeabilities were substantially larger than those predicted using the Mualem (1976) permeability function and the fitted retention curve.

(3) In the infiltration experiments the soils did not become saturated when steady conditions were achieved. Directly measured permeabilities agreed quite well with values of unsaturated permeability obtained from the pressure plate tests.

(4) A two-step method for stochastic modelling of infiltration into a liner with variable initial gravimetric moisture content, based on Loll and Moldrup (1998), was devised and shown to be feasible. The modelling incorporates the interdependence of saturated permeability and gravimetric moisture content at compaction but, at present, the retention characteristics are assumed to remain constant.

REFERENCES

Chiu, T. & C. D. Shackleford 1998. Unsaturated hydraulic conductivity of compacted sand-kaolin mixtures. *Journal of Geotechnical Engineering*, ASCE: 124(2): 160-170.

Donald, S. B. & E.A McBean 1994. Statistical analyses of compacted clay liners. Part 1: Model development. *Canadian Journal of Civil Engineering* 21: 872-882.

Fourie, A. B. & G. Papageorgiou 1995. A technique for the rapid determination of the moisture retention relation-ship and hydraulic conductivity. In E.E. Alonso & P. Delage (eds), *Proceedings of the first international conference on unsaturated soils, Unsat'95, Paris, France* 2: 485-490. Rotterdam: Balkema.

Gardner, W. R. 1956. Calculation of capillary conductivity from pressure plate outflow data. *Soil. Sci. Soc. Amer. Proc.* 20: 317-320.

Haverkamp, R., M. Vauclin, J. Touma, P.J. Wierenga, & G. Vachaud 1977. A comparison of numerical simulation models for one-dimensional infiltration. *Soil Science Society of America Journal* 41: 285-294.

Hilf, J. W. 1956. *An investigation of pore water pressure in compacted cohesive soils*. US Bureau of Reclamation, Technical Memo 654, Denver.

Kunze, R. J. & D. Kirkham 1962. Simplified accounting for membrane impedance in capillary conductivity determinations. *Soil. Sci. Soc. Amer. Proc* 26: 421-426.

Loll, P. & P. Moldrup 1998. A new two-step stochastic modelling approach: Application to water transport in a spatially variable unsaturated soil. *Water Resour. Res.* 34: 1909-1918.

Mualem, Y. 1976. Hydraulic conductivity of unsaturated soils: predictions and formulations, Methods of soil analysis. Part 1, Physical and mineralogical methods. In A. Klute (ed.), Agronomy, Am. Soc. of Agronomy, Inc., and Soil Sci. Soc. of Am., Inc., Madison, Wis.: 799-823.

Miller, E. E. & D. E. Elrick 1958. Dynamic determination of capillary conductivity extended for non-negligible membrane impedance. *Soil. Sci. Soc. Amer. Proc.:* 22: 483-486.

Remson, I., G. M. Hornberger & R. D. Molz 1971. Numerical methods in subsurface hydrology. New York: John Wiley.

Richards, L.A. 1931. Capillary conduction of liquids through porous media. *Physics* 1: 318-333.

Rogowski, A.S. 1990. Relationship of laboratory and field determined hydraulic conductivity in compacted clay layer. EPA/600/2-90/025

Shevelan, J 2000. Investigating the effects of heterogeneities on infiltration into unsaturated compacted soils. Ph.D. thesis, University of Sheffield.

van Genuchten, M. Th. 1980. A closed-form equation for predicting the hydraulic conductivity of unsaturated soils. *Soil Science Society of America Journal* 44: 892-898.

van Genuchten, M. Th., F. J. Leij, & S. R. Yates 1991. The RETC code for quantifying the hydraulic functions of unsaturated soils. EPA/600/2-91/065

Experimental Evidence and Theoretical Approaches in Unsaturated Soils, Tarantino & Mancuso (eds)
© 2000 Taylor & Francis, ISBN 90 5809 186 4

Collapse tests on isotropic and anisotropic compacted soils

M. Barrera, E. Romero, A. Lloret & A. Gens
Departamento de Ingeniería del Terreno, Cartográfica y Geofísica, Universidad Politécnica de Cataluña, Barcelona, Spain

ABSTRACT: Isotropic collapse tests under controlled matric suction have been performed on artificially prepared clay and low plasticity silt. Two different compaction processes have been selected: a one-dimensional static compaction procedure on volume basis performed on the clay and a stress-controlled isotropic static compaction procedure on the silty material. Testing programme has been accomplished with a new suction controlled triaxial cell, which allows registering the time evolution of local axial and radial deformations. A selection of results including time evolution of axial and radial strains and water content change, as well as equilibrated states at the end of every suction step, is presented for a wetting-drying cycle. A clear anisotropy upon main wetting has been observed due to the anisotropic loading condition imposed to the specimen under one-dimensional static compaction. However, this anisotropy decreases progressively upon suction induced plastic volumetric straining in the main wetting path. On the other hand, the isotropically compacted sample shows a nearly isotropic volume change evolution in the main wetting path.

1 INTRODUCTION

The paper reports isotropic stress results on two soil fabrics with collapsible tendency upon main wetting, which have been accomplished on a new suction controlled triaxial cell. Two soils have been used throughout the experimental programme, which are first described. Afterwards, the fabrication methodology is detailed, where each one of the soils undergoes a different static compaction process. These processes will induce different initial fabrics, which will display different volume change patterns of behavior upon subsequent main wetting paths. The triaxial cell is then described, followed by the testing procedures and a description of the stress paths. Finally, a selection of results is presented, where specific features of volume change behavior detected on both fabrics are discussed.

Two independent stress state variables are used: the net mean stress (σ_m-u_a) and the matric suction stress state variable (u_a-u_w), where u_a and u_w are the air and liquid phase pressures. The volume change information is represented with volumetric strain ($\varepsilon_v = \varepsilon_1 + 2\varepsilon_3$) and shear strain ($\varepsilon_s = 2(\varepsilon_1 - \varepsilon_3)/3$) evolutions, where ε_1 and ε_3 are the axial and radial strains, respectively. The information is complemented with gravimetric water content w and degree of saturation Sr evolutions.

2 MATERIAL AND COMPACTION PROCEDURES

2.1 Tested materials

Laboratory tests were conducted on two soils: a low plasticity silt from Barcelona (BCN silt) and a moderately swelling clay from Mol, Belgium (Boom clay). BCN silt has a liquid limit of

$w_L = 32\%$, a plastic limit of $w_P = 16\%$, 15% of particles less than 2 μm and a unit weight of solid particles of $\gamma_s = 26.6$ kN/m³. The dominant mineral of the clay fraction is illite, as observed from X-ray diffraction patterns (Barrera, in prep.).

Boom clay is a kaolinitic-illitic clay (20-30% kaolinite, 20-30% illite), which has a liquid limit of $w_L = 56\%$, a plastic limit of $w_P = 29\%$, 50% of particles less than 2μm and a unit weight of solid particles of $\gamma_s = 26.5$ kN/m³. Additional characteristics of the clay are described in Romero (1999).

2.2 Stress-controlled isotropic static compaction procedure

BCN silty samples were obtained following a stress-controlled isotropic static compaction procedure. Powder passing ASTM No.16-1.18 mm was left in equilibrium with the laboratory atmosphere at an average relative humidity of 47% to achieve hygroscopic water content of ca 2.2%. The silt was then sprayed with demineralised water in order to achieve predetermined water contents. A two-day curing time in sealed bags was adopted to ensure homogeneous conditions in the sample. Afterwards, the humid silt was statically compacted following a two step procedure. In the first step, a low vertical stress was applied to the soil mass confined in a three-piece split mould (35 mm in diameter and 70-mm high), until reaching a dry unit weight of 11.8 kN/m³ that was required to handle the sample. In the second step, the sample was mounted in a conventional triaxial cell under different constant net isotropic stresses (0.3, 0.6 and 1.2 MPa), which were maintained for 40 min.

Isotropic static compaction results are presented in Figure 1. Contours of equal total suction are also plotted in the figure. These contours result from interpolations of different data (ca 500 results) obtained under varying water contents and dry densities, following an equivalent compaction procedure to that previously described. Soil total suction was measured using transistor psychrometers (Woodburn et al. 1993). Careful and repeated measurements were carried out in the low suction range (lower than 200 kPa), where the psychrometers do not show so good repeatability. As observed in the vertical contour lines, no important water content change is detected upon loading at fixed suction values higher than 2 MPa. The loading mechanism affects mainly the macroporosity that does not contain free water, because moisture at these low humidity values (gravimetric water contents lower than 9%) is contained at microscale (Romero et al. 1999; Romero & Vaunat 2000, this issue).

A striking aspect is the low degrees of saturation reached after compaction at low suction values and near optimum conditions. This fact is associated with the difficulty of the experimental setup and compaction process in expelling air, despite air drainage being allowed from both bottom and top platens. Upon load application the soil particles and aggregates are brought closer by the expulsion of air. However, this process affects the subsequent expulsion of air due to it entrapment, mainly in the central zone of the specimen separated from the drainage platens. Another limitation of the technique refers to the difficulty in controlling the final volume due to end restraint effects from the drainage platens.

The initial condition of the sample prepared for isotropic testing is also indicated in the figure, corresponding to a water content of (11.0±0.2)%, a dry unit weight of 16.3 kN/m³ and a degree of saturation of ca 47%, fabricated at a net isotropic stress of 0.6 MPa. In this case, the same compaction process is followed with different initial sample geometry (50 mm in diameter and 100-mm high). The dry unit weight is determined from a trimmed sample obtained from the compacted specimen, which presents some non-uniformity due to restraints from end platens. This trimming is necessary to achieve uniform sample dimensions of 38 mm in diameter and 76-mm high, required for installation in the triaxial cell described in section 3.1. The approximate total suction is around 0.8 MPa.

2.3 One-dimensional static compaction procedure (volume control)

A one-dimensional static compaction procedure was followed on Boom clay samples until a specified final volume was achieved under constant water content. In this type of compaction, a variable static force is gradually applied to the same soil mass confined in a rigid mould by means of the movement of a piston in a strain-controlled press (travel rate 0.5 mm/min). Air-dried powder passing ASTM No.40-425 μm and with a hygroscopic humidity of ca 3.0% was thoroughly mixed with demineralised water to achieve predetermined water contents. The mixture was allowed to stand for two days for moisture equilibration. The force of compaction determined by a load cell and the corresponding piston travel were measured at different elapsed times for the same soil. Each compression path was terminated when reaching nearly saturated states, where the static compaction test was stopped before the beginning of consolidation (wet side of optimum states are impossible to be reached). This way, relationships of net vertical stress and dry unit weight at different values of moisture content can be plotted. Static compaction curves presented in Figure 2 were calculated from these relationships for specified net vertical stresses. Further experimental details are indicated in Romero (1999).

Contours of equal suction are also indicated, which were obtained following a different test procedure (Romero 1999). Total suction values higher than 3 MPa were obtained from vapor equilibrium results and transistor psychrometer readings. Matric suction values lower than 0.45 MPa were applied using air overpressure technique. At water content values less than $w = 15\%$, increasing by compaction the dry unit weight does not change significantly the initial total suction value. As matric suction values approach the saturation line, the loading mechanism affects free water and the contours of equal suction try to incline in order to converge to the limit condition $Sr \to 100\%$.

In preparing samples of Boom clay for isotropic testing, the required quantity of demineralised water to achieve a predetermined gravimetric water content of $(15.0\pm0.3)\%$ was added to the powder, previously cured at a relative humidity of 90%. After equalization, an initial total suction of approximately $\psi \approx 2.3$ MPa was achieved with an osmotic component of around 0.4 MPa measured by squeezing technique with a transistor psychrometer (Romero 1999). Wan (1996) reported an osmotic component of around 0.3 MPa calculating this component as the numerical difference between total and matric suctions and using the contact and non-contact filter paper technique. Samples 38 mm in diameter and 76-mm high were compacted in three layers with a flanged piston and using the same soil mass, resulting in a dry unit weight of (13.7 ± 0.1) kN/m^3 and an initial degree of saturation of $(44\pm2)\%$. The initial condition of the clay fabric is indicated in Figure 2. Maximum fabrication net vertical stress was around 1.2 MPa with $K_0 = 0.37$ $((\sigma_m-u_a) = 0.7$ MPa), measured by an active lateral stress system under oedometer conditions (Romero 1999). This compaction procedure was selected in order to arrive to a speci-

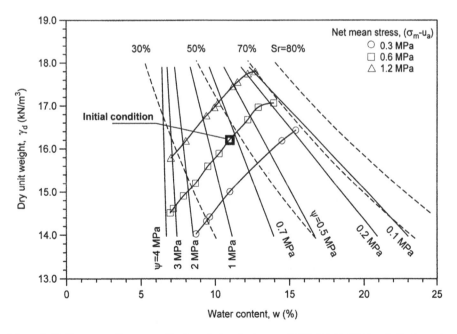

Figure 1. Static compaction curves for three isotropic stresses (BCN silt). Solid lines indicate contours of equal total suction after compaction. Initial condition of soil fabric (Barrera, in prep.).

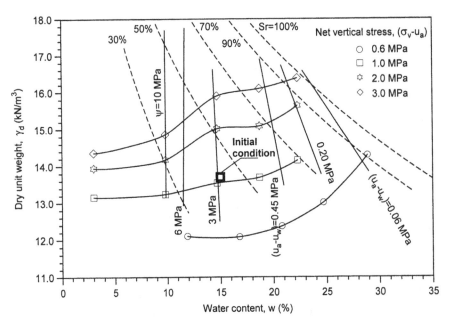

Figure 2. One-dimensional static compaction curves with contours of equal suction (Boom clay). Initial condition of soil fabric (Romero 1999).

fied volume that allowed a simpler assembly in the cell without trimming. However, a constant stress criterion using one-dimensional compaction has been preferred by Cui & Delage (1996), because it provides the same stress level in the different compacted layers, improving even more the homogeneity of the samples.

3 EQUIPMENT, EXPERIMENTAL PROCEDURES AND STRESS PATHS

3.1 Description of the triaxial cell

A triaxial cell designed to apply stress and suction paths has been built according to the cross-section scheme shown in Figure 3 and described in Romero et al. (1997) and Romero (1999). Main features of the equipment can be summarized as follows. Matric suction is applied simultaneously via air overpressure technique to both ends of the sample, maintaining a constant air pressure (number 8 in Figure 3) and controlling water pressure (numbers 9 and 10 in Figure 3) that maintains a difference equal to the prescribed suction. Both top and bottom platens include a combination of two porous stones: a peripheral annular coarse one (10-μm pore size; number 12 in Figure 3) connected to air pressure and an internal HAEV one (1.5 MPa of bubbling pressure: number 11 in Figure 3). This double drainage ensures a significantly shorter equalization stage for liquid pressure, an important advantage when testing low permeability unsaturated soils. Water content changes in the soil are calculated measuring the water volume that crosses both HAEV discs by means of two burettes with 10-mm^3 readability. These volumes are corrected taking into account the amount of air diffusing through the ceramic discs. Axial displacements are measured internally using two miniature LVDT transducers (number 2 in Figure 3) adhered to the membrane and covering the central part of the specimen. Radial deformations on two diametrically opposite sides are measured by means of an electro-optical laser system (2-μm resolution) mounted outside the chamber at a stand-off distance of 50 mm from the target surface (number 3 in Figure 3). The surface of the rubber membrane is illuminated by the coherent laser light, which crosses 15-mm transparent perspex wall (number 13 in Figure 3). A novelty of the apparatus is that the lateral measuring system can be moved up and down by means of an electric motor (number 15 in Figure 3). This way, the whole profile of the sample from pedestal to cap can be measured with the same strain resolution. In general, magnitudes of applied corrections are such that axial strains are determined with somewhat higher precision and lower bias than lateral strains. Sources of error involved in the non-contact lateral strain measurement are discussed in Romero (1999).

3.2 Experimental procedures and isotropic stress paths

Aspects of specimen mounting procedure in the triaxial cell are described in Romero (1999). In the initial stage of the tests, both cell and air pressures were simultaneously increased in steps maintaining a difference within $(\sigma_m\text{-}u_a) < 20$ kPa under liquid phase undrained conditions. This

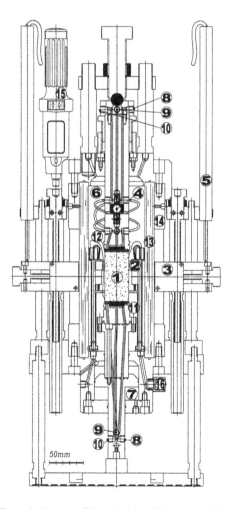

Figure 3. Layout of the triaxial cell (Romero 1999). 1) Specimen; 2) LVDT (axial strain); 3) laser displacement sensor (radial strain); 4) load cell or alignment device (isotropic test); 5) LVDT (vertical displacement of laser sliding subjection); 6) confining air/silicone oil pressure; 7) vertical stress pressure chamber; 8) air pressure; 9) water pressure (to volume change measuring system); 10) water pressure (to diffused air flushing system); 11) HAEV ceramic disc; 12) coarse porous ring; 13) perspex wall; 14) steel wall; 15) vertical displacement electric motor; 16) electrical connections to transducers and data acquisition system.

isotropic seating pressure was required to reduce bedding errors associated with LVDT fittings. In the case of the silty sample, target air pressure was selected at 0.9 MPa, whereas on the clayey sample it was maintained at 0.5 MPa. After this stage, the confining pressure of the BCN silt was raised in steps at constant $(u_a-u_w) = 0.8$ MPa up to a target $(\sigma_m-u_a) = 0.6$ MPa. Every loading step (0.05, 0.2, 0.4 and 0.6 MPa) was maintained for two days. In the case of the Boom clay sample, cell pressure was increased in steps to the desired value $(\sigma_m-u_a) = 0.6$ MPa under water-undrained conditions. Matric suction was assumed to remain approximately constant at an estimated value of $(u_a-u_w) = 1.9$ MPa, according to retention curve results presented in Romero (1999).

Wetting and drying paths, under a constant isotropic stress of $(\sigma_m-u_a) = 0.6$ MPa, were applied by varying matric suction to both soil samples. The suction paths were imposed by maintaining the air pressures previously indicated and controlling the water pressures acting on the HAEV ceramic discs. The following suction steps were applied to the BCN silt in the wetting path: 0.40, 0.30, 0.10 and 0.02 MPa. Afterwards, the silty sample was subjected to a drying path: 0.02, 0.10 and 0.15 MPa. On the other hand, Boom clay sample was tested following these suction steps in the wetting path: 0.45, 0.20, 0.06 and 0.01 MPa. Subsequently, the steps were reverted in the drying path: 0.01, 0.06, 0.20 and 0.45 MPa.

4 TEST RESULTS AND INTERPRETATIONS

4.1 Test results on BCN silt

Figure 4 shows time evolution of axial, radial, shear and volumetric strains, as well as water content and degree of saturation changes, that undergoes the BCN silty sample upon applying a suction step from 0.10 MPa to 0.02 MPa under a constant isotropic net mean stress of $(\sigma_m-u_a)=0.6$ MPa. Radial strain evolution corresponds to the central part of the specimen (indicated with subscript 'c'). Strain ratio $\varepsilon_l/\varepsilon_{3c}$ evolution is also indicated for a complementary description of sample distortion upon main wetting.

Certain tendency is observed in early stages corresponding to some swell detected by the local LVDTs, despite that the stress path has already reached the loading-collapse LC yield surface as defined by Alonso et al. (1990). At this stage the sample is loaded to the maximum fabrication stress and at a lower suction. This apparently surprising behavior may be explained in terms of certain increase in the entrapped air pressure developed during the initial stages of the transient wetting phase, which induces a small net stress unloading on the sample. This effect has also been observed by Romero (1999) when testing Boom clay in this double-drainage system. Differences between both laser 1 and laser 2 radial strain readings are due to horizontal uncontrolled displacements induced by the movement of the free cap (an alignment device is used to control the initial tilting of the sample). However, when adding both readings, the horizontal components as rigid body are compensated and the local deformation is obtained. Further aspects of lateral displacement errors are discussed in Romero (1999).

A relatively low degree of saturation is reached in the sample when applying a matric suction of 0.02 MPa. This phenomenon is partly associated with the double-drainage system used in the triaxial cell, where entrapped air in the center of the specimen is difficult to be released. Further collapse strains are expected to occur if more water is allowed to enter the soil. However, suction application with two independent air regulators, which is the system used in the experimental setup, is difficult to control at low air overpressure values. Somewhat higher degree of saturation values (ca 78% at $(u_a-u_w) = 0.05$ MPa) have been detected in oedometer cells with a bottom HAEV ceramic and following a similar wetting path with the same soil fabric (Barrera, in prep.).

As expected for the isotropically compacted specimen, a nearly isotropic evolution of axial and radial strains is observed, which is detected by the low shear strains developed and by the strain ratio $\varepsilon_l/\varepsilon_{3c}$ evolution tending to one. These results can be compared with the evolution presented in Figure 7 corresponding to the anisotropically compacted Boom clay, where a clear shear strain development is detected with $\varepsilon_l/\varepsilon_{3c} \to 0.6$.

Wetting and drying results corresponding to the different equalization stages in terms of axial, radial, shear and volumetric strains, as well as water content and degree of saturation changes, are indicated in Figure 5. Radial strains are measured in the central zone of the specimen (indicated with subscript 'c'). Shear and volumetric strains, as well as degree of saturation values, are calculated based on these mid-height radial strains. Shear strain results display a nearly isotropic evolution in both wetting and drying paths, showing some distortion of the sample in the initial wetting stages. On the other hand, no important collapse strains are detected in these first wetting stages, even though the stress path has already reached the loading-collapse LC yield surface. Similar results have been observed by Barrera (in prep.) when testing the same soil fabric in oedometer cells, where collapse strains are concentrated in the last wetting step (suction values lower than 0.15 MPa). This fact is associated with the shape of the LC yield surface in the $(u_a-u_w):(\sigma_m-u_a)$ plane, where it is expected that the highest changes in preconsolidation stresses with suction are developed at suction values lower than 0.10 MPa.

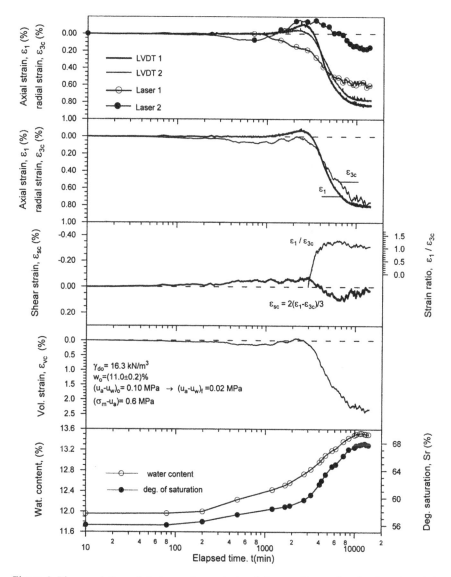

Figure 4. Time evolution of strains, water content and degree of saturation for the BCN silt in a main wetting step: $(u_a-u_w)_o = 0.10$ MPa $\rightarrow (u_a-u_w)_f = 0.02$ MPa (Barrera, in prep.).

39

The progressive development of lateral strains for the different wetting and drying steps are represented in Figure 6. The first profile identified with a value of $(u_a\text{-}u_w) = 0.80$ MPa, represents the cumulative strain of the loading path until reaching $(\sigma_m\text{-}u_a) = 0.6$ MPa. Further profiles represent the cumulative strain up to different equalization stages that are indicated with their respective target suctions. Average lateral strains are also represented in the figure and indicated with hyphenated symbols. A somewhat lower average later strain is detected for the suction step 0.10 MPa $\rightarrow 0.02$ MPa (ca 0.63%) compared to the mid-height lateral strain. Homogeneous collapse and shrinkage deformations are developed throughout the specimen height, showing the suitability of this compaction procedure compared to the non-uniformity detected in the volume control procedure (refer to Figure 8).

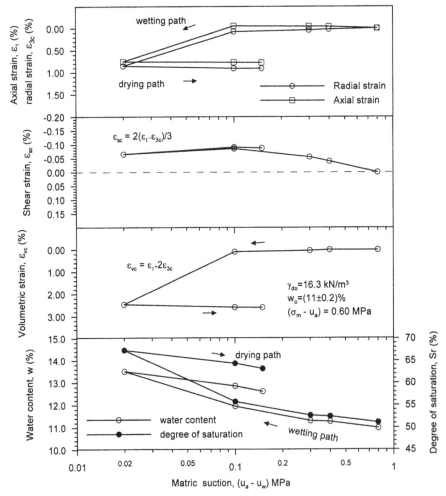

Figure 5. Variation of axial, radial, shear and volumetric strains, water content and degree of saturation for the BCN silty sample in wetting-drying cycles under $(\sigma_m\text{-}u_a) = 0.6$ MPa (Barrera, in prep.).

4.2 Test results on Boom clay

Figure 7 shows equivalent time evolution plots for the Boom clay sample upon applying a suction change starting from the initial condition $(u_a\text{-}u_w) \approx 1.9$ MPa to a final matric suction of 0.45 MPa under a constant isotropic net mean stress of $(\sigma_m\text{-}u_a) = 0.6$ MPa. Radial strain evolution corresponds to both the intermediate part of the specimen (indicated with subscript 'c' and represented with solid lines) and to the average value throughout the sample height (indicated with hyphenated symbols and represented with dashed lines).

At the beginning of the path some small swelling is recorded by the local LVDTs, before the macroscale contacts between aggregates become weaker and fail under local shear causing the macrostructural collapse. This transition is expected to occur at an estimated value of 1.4 MPa (refer also to Figure 9), when the stress path arrives to the loading-collapse LC yield locus. This yield locus is further dragged along during imbibition originating a macrostructural strain hardening of soil structure.

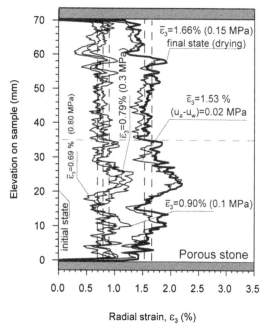

Figure 6. Cumulative radial strains throughout the BCN sample height in loading, wetting and drying paths (Barrera, in prep.).

No significant tilting of the sample is detected in Figure 7, as indicated by the small differences between both laser 1 and laser 2 radial strain readings and to both local LVDT outputs. As expected for the one-dimensionally compacted specimen, a clear shear strain development is observed in this figure upon initial wetting tending to ε_s = -0.65% and a strain ratio of $\varepsilon_l/\varepsilon_{3c}$ = 0.6.

Isochrones showing the progressive development of the lateral profiles of the sample for the different stages indicated in Figure 7 with vertical dashed lines, are represented in Figure 8. As observed, an inhomogeneous collapse deformation is developed along the specimen height as the wetting front advances, partially affected by end restraint effects of the porous stone platens tending to null deformation conditions at both ends. Maximum collapse zone is shifted downward with respect to the mid-height of the sample, condition that can be probably associated with the non-uniformity of the static compaction procedure based on a volumetric criterion rather than on a maximum stress criterion. The volume change determination taking into account the non-uniformity of the sample deformation is more representative of the whole specimen, specially when calculating degree of saturation changes. Figure 7 also represents degree of saturation, volumetric and shear strains calculated on an average radial strain basis (these last evolutions represented with dashed lines), where systematically lower collapsible strains are reported compared to results obtained from a mid-height basis due to end restraint effects. However, main differences between mid-height and average radial strains appear in this first wetting step, tending to similar values as the wetting process advances, according to results presented in Romero (1999).

Wetting and drying results corresponding to the different equalization stages in terms of axial, radial, shear and volumetric strains, as well as water content and degree of saturation changes, are indicated in Figure 9. Irreversible features affecting strains and water content variable are identified along these paths. The initial wetting affects the rearrangement and water storage capacity of the macrostructure of clay skeleton, which are not erased by the following drying path. A higher degree of saturation, compared to the value indicated in Figure 5, is reached in the sample when applying a matric suction of 0.01 MPa. However, full saturation is difficult to achieve due to air entrapment associated with the double-drainage system used in this cell. A higher degree of saturation (ca 100% at $(u_a-u_w) = 0.01$ MPa) has been detected in an oedometer cell with a bottom HAEV ceramic and following the same wetting path with the same soil fabric (Romero 1999).

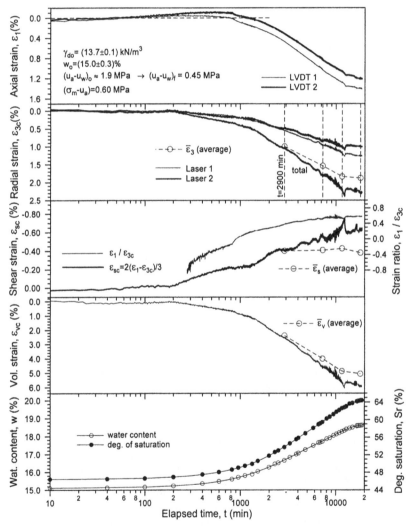

Figure 7. Time evolution of strains, water content and degree of saturation for the Boom clay fabric in a main wetting step: $(u_a-u_w)_o \approx 1.9$ MPa $\rightarrow (u_a-u_w)_f = 0.45$ MPa (Romero 1999).

Shear strain development shows distortion of the sample upon main wetting and during the activation of the LC yield locus. This effect arises from the anisotropic loading condition imposed to the specimen during one-dimensional static compaction, leading to a preferential position of clay aggregates with their larger axis set horizontally. This preferential and transversally isotropic fabric is assumed to provide the greatest resistance to applied stresses in the axial direction. This way, a lower collapse potential is expected in this direction, condition that is observed in Figure 9. Higher swelling strains in the axial direction have also been detected by Romero (1999) when testing one-dimensionally and heavily compacted samples, which displayed swelling tendency upon main wetting. This material anisotropy is progressively erased upon suction induced plastic straining at constant isotropic net stress, tending to an isotropic evolution in the following main drying path. The progressively decrease of anisotropy upon suction induced plastic straining in a main wetting path has also been observed by Romero (1999) when testing one-dimensionally and heavily compacted samples, but in this case associated with the activation of the suction decrease yield locus as defined by Gens & Alonso (1992).

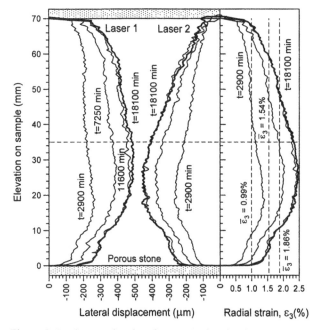

Figure 8. Isochrones showing the progressive development of the lateral profile and lateral strains of Boom clay sample in a main wetting step: $(u_a-u_w)_o \approx 1.9$ MPa $\rightarrow (u_a-u_w)_f = 0.45$ MPa (Romero 1999).

5 SUMMARY AND CONCLUSIONS

As a summary of the equipment used, the tests performed and the results obtained, the following conclusions may be drawn. The apparatus described in the present paper is suitable equipment for testing unsaturated soils under suction control with air overpressure technique. Although the double-drainage system has the advantage of reducing equalization time, a certain limitation of the experimental procedure has been detected concerning air entrapment that makes it difficult to reach high saturation degrees at low suctions. The proposed electro-optical laser measuring system has produced reliable and stable readings of lateral deformations throughout the sample height and along the different stages. Inhomogeneous collapse deformations affected by end restraint effects, as well as tilting of the sample associated with the movement of the free cap, have been clearly detected with this lateral strain measuring system.

The tests have revealed important aspects concerning compaction procedures, where a stress-controlled isotropic static compaction process has been compared with a one-dimensional static compaction procedure on volume basis.

Testing programme has provided results of collapse behavior of two different artificially prepared unsaturated soils (a low plasticity silt and a clay) under isotropic stress states. Test results show that suction decrease on a normally consolidated structure revealed irreversible collapse due to macrostructural rearrangement of soil skeleton. A clear anisotropy upon main wetting has been observed in the one-dimensionally compacted specimen. Test results also present a consistent pattern of fabric anisotropy erasure upon plastic straining in a wetting path, offering an interesting challenge for future research in constitutive modeling.

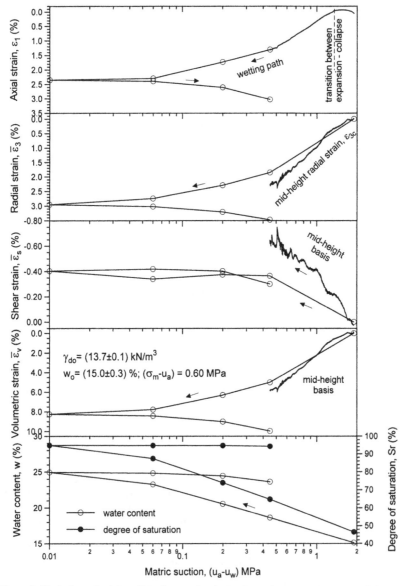

Figure 9. Variation of axial, radial, shear and volumetric strains, water content and degree of saturation for the Boom clay sample in a wetting-drying cycle at constant $(\sigma_v - u_a) = 0.6$ MPa (Romero 1999).

ACKNOWLEDGEMENTS

The first and second authors acknowledge the financial support provided by TDOC grant from the Comissionat per a Universitats i Recerca de la Generalitat de Catalunya. The first author also acknowledges the support provided through CONACYT (México).

REFERENCES

Alonso, E.E., A. Gens & A. Josa 1990. A constitutive model for partially saturated soils. *Géotechnique* 40(3): 405-430.

Barrera, M. in preparation. *Análisis teórico y experimental del comportamiento mecánico de suelos compactados*. PhD Thesis, Technical University of Catalonia, Spain.

Cui, Y.G. & P. Delage 1996. Yielding and plastic behaviour of an unsaturated compacted silt. *Géotechnique* 46(2): 291-311.

Gens, A. & E.E. Alonso 1992. A framework for the behaviour of unsaturated expansive clays. *Can. Geotech. J.* 29: 1013-1032.

Romero, E., J.A. Facio, A. Lloret, A. Gens & E.E. Alonso 1997. A new suction and temperature controlled triaxial apparatus. *Proc. 14th Int. Conf. on Soil Mechanics and Foundation Engineering, Hamburg, 6-12 September*: 185-188. Rotterdam: Balkema.

Romero, E. 1999. *Characterisation and thermo-hydro-mechanical behavior of unsaturated Boom clay: an experimental study*. PhD Thesis, Technical University of Catalonia, Spain.

Romero, E., A. Gens & A. Lloret 1999. Water permeability, water retention and microstructure of unsaturated Boom clay. *Engineering Geology* 54: 117-127.

Romero, E. & J. Vaunat, 2000. Retention curves of deformable clays. *Proc. International Workshop on Unsaturated Soils: Experimental Evidence and Theoretical Approaches, Trento, 10-12 April 2000*. This issue. Rotterdam: Balkema.

Wan, A.W.L. 1996. *The use of thermocouple psychrometers to measure in situ suctions and water contents in compacted clays*. PhD Thesis, University of Manitoba, Canada.

Woodburn, J.A., J. Holden & P. Peter 1993. The transistor psychrometer: a new instrument for measuring soil suction. In S.L. Houston and W.K. Wray (eds.), *Unsaturated Soils Geotechnical Special Publications N° 39*: 91-102. Dallas: ASCE.

Experimental Evidence and Theoretical Approaches in Unsaturated Soils, Tarantino & Mancuso (eds)
© *2000 Taylor & Francis, ISBN 90 5809 186 4*

Retention curves and 1-D behaviour of a compacted tectonised unsaturated clay

A. Di Mariano, C. Airò Farulla & C. Valore
Dipartimento di Ingegneria Strutturale e Geotecnica, Università degli Studi di Palermo, Italy

ABSTRACT: Roller-compacted tectonised clays were successfully used, in Italy, for the core of two dams built in Sicily in 1962 and are to be used for other large dams. Researches have been, as yet, carried out on this clay as a saturated material. The degree of saturation of the core of an earth-dam is lower than unity, at least before impoundment and before advancement of the wetting front. This is the reason why a great interest has arisen concerning the study of tectonised clays in the unsaturated range. First laboratory results obtained on one of this compacted tectonised nearly saturated clay are reported and discussed in this paper. Soil water retention curves and osmotic suction were determined. A new suction controlled oedometer was used to perform laboratory tests intended to simulate the deformations of the compacted clay after its loading and wetting.

1 INTRODUCTION

Highly tectonised clays have never been tested and studied in the unsaturated range even though they are non-saturated in many cases. In natural formations, for example, these clays are unsaturated up to a few meters from the ground surface. Saturation of superficial layers may then give rise to widespread shallow slides, which have not yet been fully understood (Fig. 1).

In earth embankments the compacted clay remains, almost always, in the unsaturated state except for the top layer that can become saturated after rain infiltration.

Figure 1. Shallow slides in tectonised clays frequently stud natural slopes in the winter-spring period. Photograph of the left Imera M. River, near Blufi in Sicily (the width of the picture is approximately 500 m).

The clay used for the core is unsaturated all through the construction; when the reservoir retained by the dam is being filled or emptied the wetting front within the core can advance or recede. This process can be cyclic.

It is known that both deformability and shear strength of the unsaturated clay vary with time depending on the degree of saturation and the associated suction value. Consequently there is a great interest in studying these clays in the unsaturated range.

Highly tectonised clays differ from other clays because of their peculiar structure that does influence their mechanical behaviour.

In this paper the first experimental results on a compacted highly tectonised unsaturated clay are reported.

2 PROPERTIES OF THE CLAY

Many varieties of tectonised clays exist that belong to different geological formations. They differ as to mineralogical composition, index properties and other physical properties but show two peculiar features: they are all composed of clay aggregates having the peculiar shape of "scales" and are intersected by close-spaced joints (Valore 1995). Tectonised clays are always heavily jointed, to use Rock Mechanics terminology (Hoek 1983).

The size of these aggregates ranges from a millimetre to some centimetres; they may be platy or equidimensional; their surfaces, dull or shiny, are frequently striated or slickensided. The edges of the scales are generally sharp (angular).

The clay aggregates and the structural discontinuities, other than syngenetic, derive from the distortions experienced by the sedimentary clay mass during tectonic events subsequent to its deposition. Original bedding planes and stony layers that survived the effects of the intense tectonic actions are now found folded, sheared, broken.

Within the unweathered clay, the clay aggregates are fairly well interconnected or interlocked into a three-dimensional array.

The contacts among aggregates, generally not cemented, are of the planar or concave-convex type. Each aggregate is keyed into the surrounding ones, so that its number of degrees of free-

Figure 2. Vertical section of a sample of tectonised clay compacted *in situ* by rubber-tired roller. The sample was taken from the core of Rossella dam during construction, placed into a rigid PVC container, submerged in water for 17 years and then subjected to triaxial compression. Note that clay "scales" do survive the compaction process. (Ruler scale: centimeters). (from Airò Farulla & Umiltà 1979)

dom is rather low, at low deviatoric stresses. At higher shearing stresses the edges of the aggregates and the aggregates themselves fail, creating new smaller fragments. Increments of the compressive and shearing stresses bring about the damaging of the aggregates, which however do survive up to very high stress levels and to compaction process (Airò Farulla & Umiltà 1979; Valore, 1991; Airò Farulla & Valore 1993).

These clays are often designated as "clay-shales" or "shales". However they are not always indurated nor have "the property of splitting easily along closely spaced parallel planes" (Twenhofel 1937). The terms "tectonised" or "scaly" clays seem more appropriate.

2.1 The in situ compacted clay

Owing to their structural features and natural water content these clays may be placed and compacted as excavated to obtain a material with low permeability and appreciable shear strength and stiffness, suitable as a core material in an earth dam.

In Italy roller-compacted tectonised clays were used, successfully, for the core of two dams built in Sicily in 1962, namely Scanzano and Rossella dams, 40 and 30 m high respectively and are to be used for other large dams. Among these a 73 m high embankment dam to be built near Blufi in Sicily.

In the cores of Scanzano and Rossella dams, as well as in the Blufi test fill, a compact arrangement of the scales was generally obtained. The material is formed predominantly by sharp-edged clay fragments with sizes in the range 1-10 mm, closely packed together; between the scales a thin clay "flour" is frequently interposed. Rare chunks of clay (3-15 cm) survive the compaction process, see e.g. Figure 2.

The roller compaction leads to a non-ordered structural arrangement of the clay fragments, definitely random as a whole, except near the interfaces of contiguous layers.

The interface between successive lifts is evident and rather regular (Airò Farulla & Valore 1993).

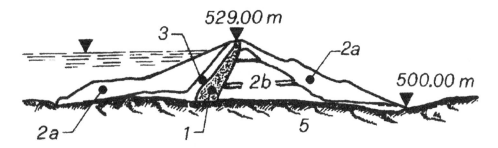

Figure 3. Schematic maximum section of Rossella dam. 1 stiff tectonised clay; 2a sandy gravel; 2b sand; 3 sand and crushed soft sandstone transition; 5 foundation soil: stiff tectonised clays. Elevations in m above mean sea level. (Airò Farulla & Valore 1993)

2.2 Physical characteristic of Rossella clay

The clay used for the experiments on the unsaturated material was taken from the core of Rossella dam. The cross section of this dam is shown in Figure 3. It is a kaolinitic and illitic clay having typically a liquid limit, w_l, of 58%, a plastic limit, w_p, of 28% and a clay fraction ($<2\mu m$) of 49% and specific weight, γ_s, equal to 27.3 kN/m^3.

Some other data are reported by Airò Farulla & Valore (1993).

2.3 Preparation of the laboratory compacted clay samples.

For a better control over the effects of water content and suction of the material during the experiments, the clay - taken from the uppermost layers of the core in 1998 - was disaggregated and dynamically re-compacted in the laboratory. In this way a quite homogeneous material was also obtained.

The samples taken from the core were disaggregated with the use of a rubber hammer in order to avoid the complete disintegration of the clay aggregates, which seem to greatly influence the mechanical behaviour of the material. The disaggregated clay was passed through ASTM sieve No 4 and was dried, for a minimum of 48 hours, at a temperature of 60°C, in order not to alter both the diffuse-layer and the strongly bounded water.

The dry clay was then mixed with distilled water to obtain a sample having a given water content. The prepared material was put inside three plastic bags hermetically closed and placed in a room with a relative humidity greater than 85%. A uniform distribution of moisture in the sample was thus presumably achieved.

The compaction was obtained with the Standard Proctor method, doubling the energy of compaction (1.2 MJ/m^3). This is the method that best seems to reproduce the *in situ* condition of the clay in the core of the dam. Only one layer of the material was used in the compaction procedure.

3 RESEARCH PROGRAMME

The research programme to be carried out on unsaturated Rossella clay includes the determination of the retention curve, osmotic suction, 1-D deformability, of course, swelling pressure, swelling behaviour in relation to suction and its shear strength. Compaction effects on the structure of the clay and, in general, on its mechanical behaviour are being investigated.

Up to now the retention curve, the osmotic suction and 1-D deformability of the clay have been studied. The research is still in progress.

4 DETERMINATION OF THE RETENTION CURVE

The retention curve expresses the relationship between the water content of a soil and its suction. It is also called soil moisture characteristic curve or simply characteristic curve.

The determination of the retention curve is of primary importance in the study of the mechanical behaviour of unsaturated soils; in the study of the deformability of clays used as core materials in embankment dams it plays a crucial role.

In a clayey material of low porosity, like the one used to build the core of Rossella dam, suction can have high values (higher than 1 MPa) even when the degree of saturation is greater than 85%. Hence, the choice of the instrument to use for the determination of the characteristic curve is limited, at least in the high suction range.

The retention curve of Rossella clay was obtained with the use of a transistor psychrometer, in the range of suction from 1 to 30 MPa, and with an axis translation technique suction controlled oedometer, in the range 0.01-1 MPa.

4.1 *Transistor psychrometer*

The SMI transistor psychrometer (Woodburn et al. 1993) allows the measurement of the relative humidity of the air within a confined space. The transistors are extremely sensitive to very small temperature changes and this fact enables them to measure relative humidities above 92% and as high as 99.85%. The transistor psychrometer can thus be used to measure total soil suction values from 0.2 MPa to greater than 10 MPa.

It is known that the accuracy of the instrument is greater when the measurement conditions are as close as possible to the calibration ones, therefore, the psychrometer was calibrated before starting the tests.

The probes of the psychrometer were calibrated using standard sodium chloride solutions having different concentrations. Equilibrium vapour pressures of each solution are known and therefore corresponding total suctions are also known.

Concentrations of salt solutions used (at the laboratory temperature of 20° C) and corresponding total suctions Ψ in MPa were obtained with the following expression (Romero 1999):

$$\Psi = -135.1 \times \ln\left[1 - 0.035m - 1.1421 \times 10^{-3} m(m-3)\right]$$

where m is the salt solution molality in mol/kg. This expression is only valid for values of m not greater than 3.0 mol/kg

4.2 Specimen preparation and experimental procedure

Six specimens were extracted from each compacted sample with a cylindrical sampler having an internal diameter of 15 mm. All six specimens were used to determine a single retention curve in order to attempt minimizing the measurement errors as well as the possible influence of local differences (within the same sample) in water content and structure on the retention curve.

All of the specimens were prepared under temperature controlled conditions. When extracted from the sampler, the soil was put inside the PVC long sampling tubes of the psychrometer and then accurately cut to a height of 12 mm.

The weight of each specimen was determined before starting the test. All the specimens were weighed with an electronic balance having a resolution of 0.0001g.

The main drying retention curve was obtained with a step by step procedure. Each suction measurement lasted one hour. Every single specimen was weighed at the beginning and at the end of each measurement and the zero of the psychrometer was reset after every single suction determination. The test finished when the suction of each specimen reached the upper limit of the instrument. On average, sixteen suction measurements were made for each specimen.

At the end of a test, the final gravimetric water contents of the specimens were determined. With these values, the water contents corresponding to the measured suctions were calculated through a backward procedure. The water retention curve, for a given initial dry unit weight and water content of the clay, was thus determined.

Volume variation of each specimen, as it was verified, can be considered negligible throughout the entire drying process.

4.3 Experimental results and interpretation

Water retention curves were determined with the transistor psychrometer for three samples having different initial water content values and dry unit weights. These samples were used to determine the compaction curve of the clay, at constant compaction energy.

The three samples used to determine the retention curves were compacted on the dry side of the compaction curve (optimum gravimetric water content close to 18%). Only the drying paths of the retention curves were determined.

Water retention curves are shown in Figure 4. The three curves, starting from different initial water contents and following different paths, converge to the same curve for water contents lower or equal to about 12%.

Retention curves of Rossella clay, Proctor compacted at a double energy and having a water content lower than 12%, always coalesce into the curve shown in Figure 4 (at least for drying paths), regardless of the initial value of dry unit weight.

That means that below a water content value of approximately 12%, the macro-structure of Rossella clay has a negligible influence, if any, on its retention curve.

The data of the retention curve relative to an initial dry unit weight of 17.36 kN/m^3 (closer to that in the core of the dam) were integrated, in the range of suction 0.01-1 MPa, with the data obtained from an air-overpressure suction controlled oedometer test. This test allows to control soil matrix suction with a technique which is equal to the axis translation technique except for the fact that, in this case, air pressure is always maintained constant. The results obtained are shown in Figure 5.

It must be pointed out that the specimen in the oedometer was wetted under constant load showing swelling. This explains why the wetting path of the retention curve is not lower than the drying path as it always is the case.

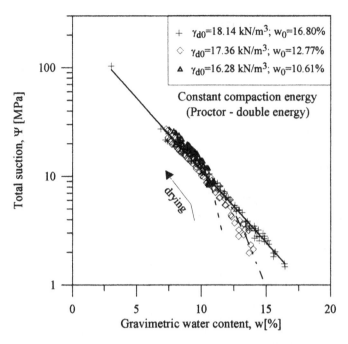

Figure 4. Water retention curves determined with the transistor psychrometer (drying paths) (Di Mariano, in prep.)

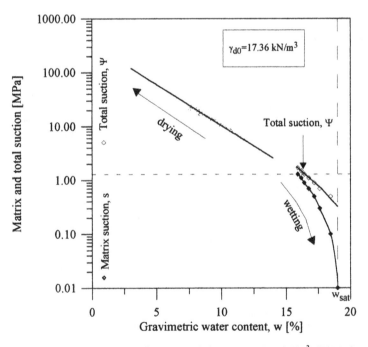

Figure 5. Water retention curve relative to γ_{d0}=17.36 kN/m^3. (Di Mariano, in prep.)

In Figure 5 experimental data obtained with the oedometer, referring to the matrix suction, are also plotted. The curve representing total suction versus water content in the low range of suction was obtained after the determination of the osmotic suction as described in the following paragraph.

5 DETERMINATION OF OSMOTIC SUCTION

Total suction is defined as the sum of matrix and osmotic suctions. Osmotic suction depends on the dissolved salts contained in the pore fluid and the mineralogical composition of the soil strongly affects its value.

The mineralogical composition of Rossella clay is not known. In order to quantify the difference between total and matrix suction in this clay and be able to calculate total suction values from the matrix suction ones, the value of the osmotic suction was determined experimentally.

The pore fluid squeezing technique was used to extract pore water from the clay. This technique consists in squeezing a soil specimen with a press to extract the macro-pores water and then measuring its electrical conductivity. The latter depends on the dissolved salts present in the water and is related to the osmotic suction.

This technique has proven to give the most reliable measurements of osmotic suction (Krahn & Fredlund 1972). In addition, Romero (1999) has compared electrical conductivity results to osmotic suction measurements obtained with the contact non-contact filter paper technique. The measured electrical conductivity values showed to closely agree with the total minus matrix suction values determined with the filter paper technique.

5.1 *Pore fluid squeezer*

The pore fluid squeezer is a heavy-walled stainless steel cylinder with a piston squeezer. In the centre of the upper part of the piston there is a concave and hemispherical seat for a high resistance steel sphere, which allows centring the vertical applied load. Three o-rings guarantee the impermeability of the system.

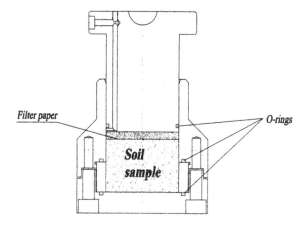

Figure 6. Pore fluid squeezer. (Di Mariano, in prep.)

The vertical load is applied to the piston through a hydraulic press. In order to maintain a constant pressure (pressure controlled system) on the piston squeezer during the test, the hydraulic circuit of the press was modified. A deviation was created that could disconnect the cylinder of the press from its motorised group and connect it to an air compressor. In the hydraulic circuit of the press, where oil is present, it is impossible to let compressed air circulate. Hence, an air-oil interface was created to convert the pressure of the air compressor into pressure on the oil present in the circuit of the press (Fig. 7).

5.2 *Specimen preparation and experimental procedure*

The soil specimen was prepared with a gravimetric water content equal to 1.6 times the plasticity limit. The water content has to be high enough to allow the extraction of an amount of pore water sufficient to measure its electrical conductivity but, at the same time, it must not be greater than the liquid limit. In the latter case, in fact, leaching of dissolved salts could take place with the consequent alteration of the original saline content of the pore water, (ASTM D4542 1993).

Before placing the soil specimen inside the pore water squeezer, every single part of the apparatus was first washed with alcohol and then rinsed once with tap water and twice with distilled water. All of this done to reduce contamination of the pore liquid as much as possible.

The specimen was put inside the pore water squeezer and, before closing the system with the piston, a filter paper and a high porosity brass plate (85 μm pore size) were interposed between the specimen and the piston.

The filter paper used was a Whatman 42 ashless filter paper, 55 mm in diameter. Before use, the paper was accurately washed with distilled water and then dried at a temperature of 105°C for 24 hours. During the test the filter paper is used solely to protect the porous plate from the clay particles. It is of great importance that the filter paper, while protecting the porous plate, does not alter the chemical content of the pore liquid. The fact that the high temperature could somewhat modify the physical characteristics of the filter paper does not significantly affect the squeezing process.

The pore liquid is extracted by means of a sterile syringe connected, by a 4 mm diameter polyamide tube, to the effluent passage in the piston squeezer.

The pore water extraction process is quite slow and the duration of each test was approximately three weeks. The syringe remains connected to the piston (with the 4 mm diameter tube) all throughout the test. A depression is applied inside the pore water squeezer by means of the syringe, in order to make the water extraction easier. Every two or three millilitres of pore water extracted, the syringe is emptied in a sterile glass container, which is kept at a temperature of 1-5°C to reduce bacteria growth in the pore liquid.

Pore water extracted at different pressures is collected in separate containers.

Before measuring the electrical conductivity, the water collected is filtered with a ultra-fine Whatman GF/C glass fibre filter paper whose pores diameter ranges between 1 and 10 μm. Filtration is necessary to eliminate flocculated clay particles that could alter conductivity readings (Romero 1999).

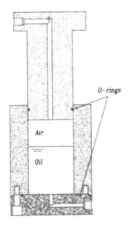

Figure 7. Air-oil interface deviced to guarantee constant squeezing pressure. (Di Mariano, in prep.)

An electrical conductivity probe was used to measure the conductivity of pore water. The apparatus used was able to automatically correct the value of conductivity with the temperature.

5.3 *Experimental results and interpretation*

The salt content of the pore water, and hence its electrical conductivity, are related to the osmotic suction of a soil. The electrical conductivity of the pore liquid extracted appears to be affected by the magnitude of the extraction pressure. Pore water was extracted applying three different pressures (2 MPa, 5 MPa and 23.5 MPa).

Relationship between electrical conductivity and osmotic suction follows an exponential law. The osmotic pressures of the extracted pore water was determined by means of the following expressions:

$$\pi = EC^{1.074} \times 0.0191 \qquad \text{(U.S.D.A. 1950)}$$

$$\pi = EC^{1.065} \times 0.0240 \qquad \text{(Romero 1999)}$$

where π is the osmotic pressure in kPa and EC the electrical conductivity in μS/cm.

The first curve was obtained with water solutions containing mixtures of dissolved salts and was proposed in the U.S.D.A. Agricultural Handbook No. 60 (1950); the second one was determined with homoionic NaCl solutions by Romero (1999). The two curves lead to quite close osmotic suction values.

Variation of the electrical conductivity of the pore water extracted versus the extraction pressure is shown in Figure 8.

When using the squeezing technique, almost all the salts contained in the macro-porosity water are extracted at the beginning of the test. As the test continues and the squeezing pressure is increased, the salts contained in the micro-porosity water will not be dissolved in the "free" water due to the "restrictive membrane effects", which do not allow the cations to freely diffuse into the macro-porosity water (Mitchell 1993). Thus electrical conductivity of the pore water extracted decreases (due to salt concentration decrease) as the squeezing pressure increases.

It appears that the value of electrical conductivity that best represents the osmotic suction of Rossella clay is the one related to the highest salt content of the extracted pore liquid.

The osmotic suction of Rossella clay calculated according to the above procedure is equal to 0.43 MPa. This value corresponds to the average osmotic suction between the values obtained with the two above expressions for π.

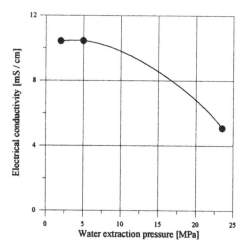

Figure 8. Electrical conductivity *vs* pore water extraction pressure. (Di Mariano, in prep.)

6 DEFORMATION UNDER SUCTION CONTROLLED CONDITIONS

The aim of these oedometer tests is to duplicate and reproduce the *in situ* conditions of the clay in the core of the dam. The intent is to simulate the deformations of the soil during and after the construction of the dam.

Stress paths for the oedometer tests were specifically chosen in order to load soil samples under constant suction and then slowly wet them.

6.1 *Suction controlled oedometer*

Deformation under suction controlled conditions was investigated by means of a new oedometer (Fig. 9) designed and constructed at the Geotechnical laboratory of the U.P.C. (Technical University of Catalonia, Barcelona, Spain).

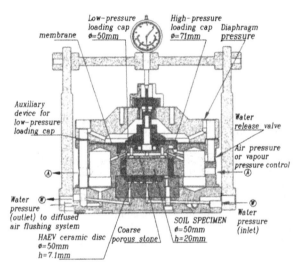

Figure 9. Schematic vertical section of the new suction controlled oedometer. (Di Mariano, in prep.)

The new apparatus allows performing oedometer tests on unsaturated soil samples controlling either matrix or total suctions, during the whole test.

Matrix suction is controlled through the application of the axis translation technique. The soil specimen rests on a high air entry value ceramic disc (1.5 MPa), inside a cell where a controlled air pressure can be applied. The ceramic disc allows the control of pore water pressure in the soil. Vertical load is applied with compressed air. Two different pistons can be used with the oedometer. One allows the application of a net vertical stress on the specimen that is equal to the difference between piston pressure and air pressure. The other allows to double the stress acting on the specimen without changing the pressure on the piston. A 1 mm thick rubber membrane separates the loading chamber from the pore air pressure one. The oedometer is all made of stainless steel while the two loading pistons are made of brass. Water volume changes are measured by means of a graduated glass burette (resolution of 0.02 mL) enclosed in a Perspex cylinder. Inside this system two immiscible liquids are present (de-aired water and a yellow coloured liquid). The meniscus formed by the two liquids moves inside the burette, which is directly connected to the specimen. Burette readings do not depend on water pressure.

Control of room temperature within a range of ±1°C, during oedometer tests, is of extreme importance to obtain accurate readings of water volume change.

Axial displacements are measured by means of a mechanical micrometer having a resolution of 2 μm.

In the new apparatus, if required, it is also possible to perform oedometer tests subjecting the specimen to a prefixed relative humidity. Control of the total suction can thus be guaranteed. At the present stage of the research it seems anyhow more interesting to investigate firstly the effects of matrix suction variations on the deformability of the soil. No control on the effects of the relative humidity surrounding the specimen has yet been tested.

6.2 *Specimen preparation and experimental procedure*

Three samples were dynamically compacted at the same initial dry unit weight in order to compare results of different tests. The initial degree of saturation of the samples was always less than or equal to 90%. This value of the degree of saturation approximately corresponds to the air entry value of Rossella clay at the initial dry unit weight considered.

Two specimens were extracted from each compacted sample: the first for the transistor psychrometer and the second for the oedometer. The former was used to measure the initial total suction of the sample. Matrix suction was then estimated subtracting the osmotic suction from the total suction. A value of ca. 1.3 MPa was found.

The other specimen, with a diameter of 50 mm and a height of 20 mm, was placed inside an oedometric ring, 6 mm thick, and then in the suction controlled oedometer. The specimen was then slowly loaded to the final value of total net vertical stress and, while loaded, its water content did not vary. At the end of the loading path, the initial suction value of 1.3 MPa was applied through axis translation technique and the soil was allowed to equilibrate with its boundary conditions. Once equilibrium was attained the specimen was wetted under constant net vertical stress, maintaining air pressure always constant at 1.4 MPa and slowly increasing water pressure (air-overpressure technique). It is important to indicate that, while increasing air pressure to the target value, some compression of the soil skeleton was detected mostly associated to pore fluid and occluded air bubbles volume changes. Further details are reported in Di Mariano (in prep.).

Three oedometric tests have been performed (the third one is still running); the stress paths followed are shown in Figure 10.

6.3 *Experimental results and interpretation*

Figures 11 and 12 show variations of water content and volumetric strain with time, during two oedometer tests (A1-B1-C1 and A2-B2-C2 in Fig. 10). When the soil was loaded and the air pressure increased, void ratio, degree of saturation and suction somehow changed. More precisely, the degree of saturation increased reaching nearly 100% and the suction decreased below

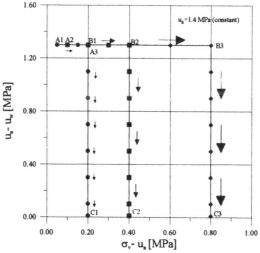

Figure 10. Stress paths followed during oedometric tests. (Di Mariano, in prep.)

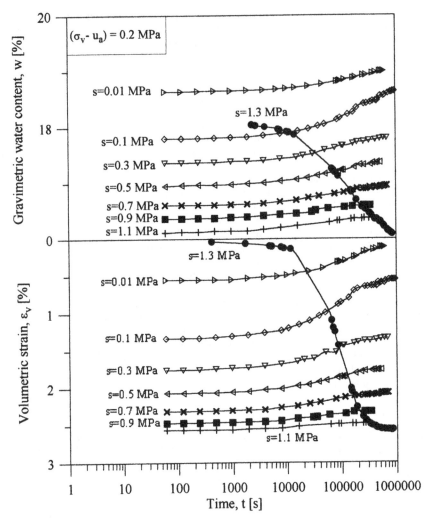

Figure 11. Gravimetric water content and volumetric strain *vs* time. First oedometer test. (Di Mariano, in prep.)

the initial value of 1.3 MPa. As soon as net vertical stress reached its final value (0.2 MPa and 0.4 MPa respectively), the sample was left to equilibrate with its initial matrix suction. During this step of the test the soil underwent shrinkage and its water content decreased (curve at s=1.3 MPa, Figs 11 and 12) until equilibrium with the boundary conditions was reached.

As matrix suction decreased (i.e., water pressure increased) the sample started to swell and its water content began to increase (Figs 11 and 12).

Swelling was rather limited in the first wetting steps and increased notably at low suction values (Figs 11 and 12).

As matrix suction decreased, the degree of saturation of the sample and its gravimetric water content increased (Fig. 13). In Figure 13, the points corresponding to the equilibrium of the specimen with its boundary conditions are plotted.

In the oedometer tests, the higher the stress on the sample the lower the swelling during the wetting path. This is shown in Figure 14 for test 1 (path A1-B1-C1 in Fig. 10) and 2 (path A2-B2-C2 in Fig. 10). The third test has not been completed yet.

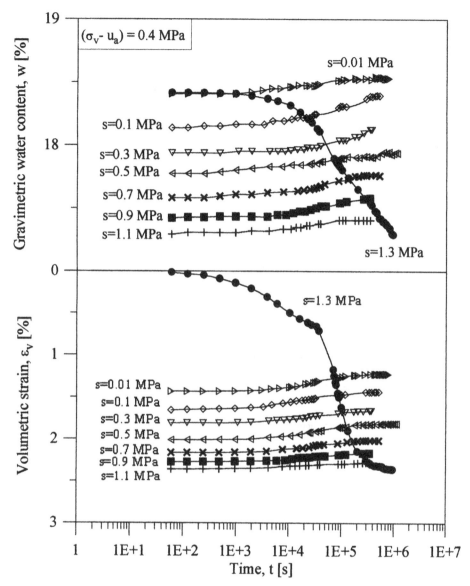

Figure 12. Gravimetric water content and volumetric strain *vs* time. Second oedometer test. (Di Mariano, in prep.)

Values of in-flowing and out-flowing volume of water through the high air-entry value ceramic disc were also used to draw the retention curve of each sample in the range 0.01-1.3 MPa of matrix suction. Results in terms of gravimetric water content and degree of saturation are shown in Figures 15 and 16. The sample with the highest void ratio was subjected to a net vertical stress of 0.4 MPa and its swelling upon wetting was lower than that of the first oedometer sample (Fig. 14). This is why the two retention curves cross.

During each oedometer test, measurements of water permeability were also performed. For every single step of matrix suction decrease applied to the soil sample, under a constant net vertical stress, transient inflow of water was measured with time. Results were interpreted with a simplified solution of Richards's equation that takes into account a non-negligible ceramic disc impedance (Kunze & Kirham 1962; Romero et al. 2000).

From Figure 17 it can be seen how the coefficient of water permeability does show a tendency to increase with the degree of saturation; however its variations are not marked within the explored range of the degree of saturation. As the degree of saturation approaches 100%, the curves in Figure 18 tend to the average permeability value of the clay taken from the core of Rossella dam.

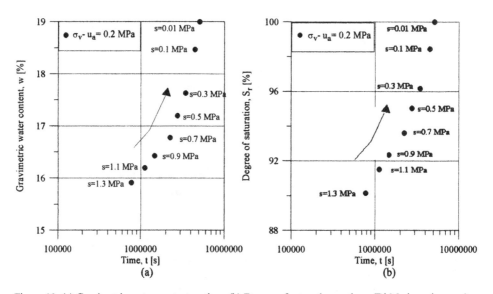

Figure 13. (a) Gravimetric water content *vs* time; (b) Degree of saturation *vs* time. (Di Mariano, in prep.)

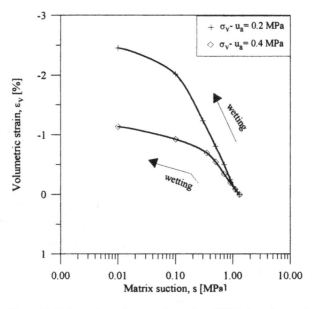

Figure 14. Volumetric strain *vs* matrix suction. (Di Mariano, in prep.)

Results shown in Figure 17 are not in contrast with the ones obtained on another highly tectonised compacted clay, Blufi clay, which are shown in Figure 18.

7 CONCLUSIONS

Highly tectonised clays were used and are to be used for the core of zoned dams in Sicily. These clays are unsaturated after completion of the dam and may become saturated after the filling of the reservoir. An understanding of the mechanical response of these soils upon wetting is therefore essential to the design for the interpretation and the prediction of the behaviour of existing and of new dams.

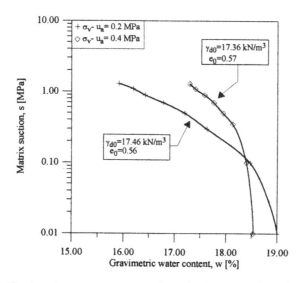

Figure 15. Gravimetric water content *vs* matrix suction for test 1 and 2. (Di Mariano, in prep.)

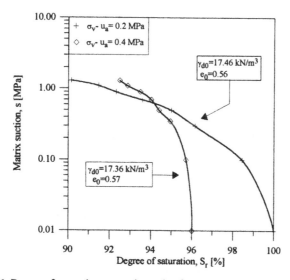

Figure 16. Degree of saturation *vs* matrix suction for test 1 and 2. (Di Mariano, in prep.)

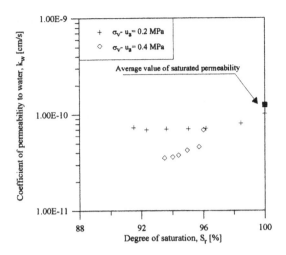

Figure 17. Variation of water permeability with the degree of saturation, during test 1 and 2. (Di Mariano, in prep.)

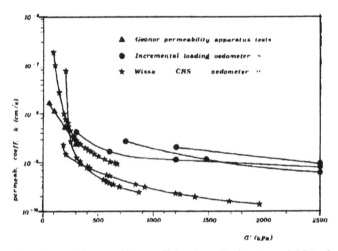

Figure 18. Variation of saturated permeability coefficient k vs effective stress σ' (kPa) of compacted saturated Blufi clay. (Valore 1991)

An experimental programme is being carried out to investigate the behaviour of unsaturated highly tectonised clays and the first experimental results are presented in this paper.

Retention curves were determined with a transistor psychrometer in the high range of suction and with an air overpressure suction-controlled oedometer in the low range. Only drying retention curves were obtained with the psychrometer and only wetting curves with the oedometer. Research is still in progress.

Retention curves determined with the psychrometer are expressed in terms of total suction whereas the curves obtained with the oedometer are in terms of matrix suction. In order to compare the results, the osmotic suction of the clay was determined by means of the pore water squeezing technique. A decrease in the electrical conductivity of the pore liquid extracted with the increase of the squeezing pressure was detected.

Deformability of the unsaturated compacted material is being studied through a number of experimental tests performed with a new suction controlled oedometer. This new apparatus al-

62

lows the application of matrix suction *via* the axis translation technique and total suction *via* vapour equilibrium technique. Results of the first tests are presented in this paper.

The partially saturated highly tectonised clay, compacted at a water content very close to its optimum and loaded to a given value of the net vertical stress, does not show a tendency to collapse as wetted, at least not for net vertical stresses lower or equal to 0.8 MPa. The clay, right after the loading path, undergoes compression and then starts swelling as wetted. Its swelling decreases as the net vertical stress increases. Loading followed by drying as well as the role played by the size of the "scales" and of the "scaly" nature of the material itself need to be studied yet.

ACKNOWLEDGEMENTS

Experimental research on the unsaturated clay was and still is carried out at the Geotechnical Engineering Department of the Technical University of Catalonia, Barcelona, Spain. The authors are greatly indebted to Prof. E. Alonso Pérez de Agreda and to all his staff.

Special thanks go to Dr. Enrique Romero and Prof. Antonio Lloret for their suggestions, guidance and support; and to Tomás Pérez for the schematic drawings of the instruments used.

REFERENCES

Airò Farulla, C. & G. Umiltà 1979. Proprietà fisico-meccaniche di argille a scaglie. Esperienze su campioni prelevati dal nucleo di una diga e sottoposti a prolungata immersione in acqua. *Rapporto di ricerca, Istituto Ingegneria Geotecnica e Mineraria, Università di Palermo.*

Airò Farulla, C. & C. Valore 1993. Some aspects of the mechanical behaviour of compacted tectonised clays. *Proc. Int. Symp. on Hard Soils-Soft Rocks, Athens*: 335-342. Rotterdam: Balkema

ASTM 1993. *Annual book of ASTM standards.* 04.08, Philadelphia.

Di Mariano, A. in preparation. *Le argille a scaglie e il ruolo della suzione sulla loro deformabilità.* PhD Thesis, Università degli Studi di Palermo, Palermo, Italy.

Fredlund, D.G. & H. Rahardjo 1993. *Soil Mechanics for Unsaturated Soils.* New York: John Wiley and Sons Inc..

Hoek E. 1983. Strength of jointed rock masses. *Géotechnique* 33 (3): 187-225.

Krahn, J. & D.G. Fredlund 1972. On total, matric and osmotic suction. *Soil Science* 114(5): 339-348.

Kunze, R.J. & D. Kirkham (1962). Simplified accounting for membrane impedance in capillary conductivity determinations. *Soil Sci. Soc. Am. Proc.*, 26: 421-426.

Mitchell, J.K. 1993. *Fundamentals of soil behavior.* Second edition, John Wiley & Sons, Inc.

Romero, E., A. Gens & A. Lloret 2000. Temperature effects on water retention and water permeability of an unsaturated clay. H. Rahardjo, D.G. Toll & E.C. Leong (eds.). Proc. of the *Asian Conf. On Unsaturated Soils, Singapore*, 18-19 May 2000: 433-438. Rotterdam: Balkema.

Romero, E. 1999. *Characterisation and thermo-hydro-mechanical behaviour of unsaturated Boom clay: an experimental study.* Tesis Doctoral, Universitat Politècnica de Catalunya, Barcelona.

Twenhofel, W.H. 1937. Terminology of the fine-grained mechanical sediments. *National Research Council*, Washington.

U.S.D.A. Agricultural Handbook No. 60. *Diagnosis and improvement of saline and alkali soils.* 1950.

Valore, C. 1991. A tectonised variegated clay as core material. *Proc. XVII Congress on Large Dams, Vienna*, Question 67, Report 18: 299-318.

Valore, C. 1995. Structure and mechanical behaviour of tectonised clays. *Proc. XI ECSMFE, Copenhagen*: 7.149-156.

Woodburn, J.A., J. Holden & P. Peter 1993. The transistor psychrometer: a new instrument for measuring soil suction. In S.L. Houston and W.K. Wray (eds.), *Unsaturated Soils - Geotechnical Special Publications N° 39, ASCE, Dallas*: 91-102.

Experimental Evidence and Theoretical Approaches in Unsaturated Soils, Tarantino & Mancuso (eds)
© *2000 Taylor & Francis, ISBN 90 5809 186 4*

Volume change behaviour of a dense compacted swelling clay under stress and suction changes

M. Yahia-Aïssa, P. Delage & Y. J. Cui
Ecole Nationale des Ponts et Chaussées, CERMES, Paris, France

ABSTRACT: The paper presents some investigations carried out on a heavily compacted swelling FoCa7 clay ($\rho_{di} = 1.9$ Mg/m^3). Tests were run on a high pressure oedometer with double lever arm, developed in the laboratory. The device is able to multiply the load placed on it by a ratio of 50. The maximum possible applied stress is equal to 30 MPa on a 70 mm diameter ring. The suction is controlled by circulating air at a controlled relative humidity at the basis on the sample. The relative humidity is controlled using saturated salt solutions. Tests in which the sample is submitted to suction cycles under constant loads were carried out. The results obtained showed a reversible response in volume, confirming previous results obtained under a zero stress. This particular behaviour is related to the high density of the samples tested, and to the major influence of physico-chemical clay-water interactions occurring at a microscopical level. In these dense samples, due to the very high compaction stress (60 MPa), no inter-aggregate pores are believed to exist, and therefore no irreversible collapse can take place, explaining the reversible response obtained.

1 INTRODUCTION

In the frame of the French research program on radioactive waste storage at great depth, various studies have been carried out on the hydro-mechanical behaviour of a heavily compacted swelling clay named FoCa7 (Atabek et al. 1991), considered as a possible engineered barrier. This clay is compacted under a high stress (60 MPa), giving a high dry density ($\rho_{di} = 1.85$ Mg/m^3). Former investigations on the determination of the water retention curve of FoCa7 clay evidenced some particular properties of this clay. No hysteresis under cycles in suction were observed, the samples giving reversible response in terms of water content and volume changes. This behaviour was related to the high density of the sample, and to the major effect of physico-chemical clay-water interactions (Delage et al. 1998).

Various workers addressed the problem of cyclic swelling and shrinkage of clays (Chen 1975, Escario & Saez 1973, Chen et al. 1985, Habib & Karube 1993, Alonso et al. 1995, Gehling et al. 1995, Romero et al. 1995, Vilar 1995a,b, Belanteur et al. 1997). Most often, the densities of the clay tested were in the range of 14-16 Mg/m^3, smaller than the FoCa7 clay density considered here. At these lower densities, the volume changes of the samples submitted to suction cycles under various constant loads correspond to a combination of swelling and collapse, depending on the number of cycles applied, and on the magnitude of the constant load supported. An interpretation can be made using the framework provided by Gens & Alonso (1992), who considered an aggregated microstructure, with an intra-aggregate and an inter-aggregate porosity. Reversible responses are given at the microstructure level, which involve intra-aggregate phenomena. Any swelling is related to the swelling of the aggregates, whereas any collapse is irreversible, and related to the collapse of the inter-aggregate pores. As commented before, cycles in suction under a zero stress performed by Delage et al. (1998) on the denser FoCa7 clay showed no irreversibility, indicating a major influence of the microscopic level, which is believed to be reversible in nature. In other words, the high stress compaction carried out while preparing the FoCa7 samples probably collapsed the pores which are responsible for the collapse observed in other looser compacted clays.

In this paper, tests in which samples of FoCa7 clay are submitted to cycles in suction under a constant load are presented, in order to extend previous conclusions related to reversibility, drawn from tests performed under a zero stress.

2 MATERIALS AND METHODS

The FoCa7 clay used in this study comes from a place named Fourges - Cahaigne, located in the North -West of France. The FoCa7 clay is composed of 80% interstratified kaolinite – smectite, 7% kaolinite and 2% quartz, together with some traces of goethite and gibbsite (Atabek et al. 1991, Sayouri 1996). The plasticity index and the specific surface of the clay are respectively equal to 62 (Atabek et al. 1991), and 515 m^2/g, according to methylene blue measurements.

Samples were statically compacted (K_0 condition) at a 0.1 mm/min compression rate, under a 60 MPa compaction stress, from a clay powder at a water content w = 12.5 ± 0.5%. After compaction, specimens were placed in dessiccators containing a saturated K_2CO_3 solution, imposing a suction value approximately equal to 113 MPa. The average dry density and degree of saturation obtained were respectively ρ_d = 1.9 Mg/m^3 and S_r = 80%.

High stresses were applied to the samples by using the high stress oedometer frame described in Figure 1-a. With a double lever arm, a ratio of 50 is applied to the load placed on the device, leading to a maximum force of 12 tons, corresponding to a stress of 30 MPa with a 70 mm diameter ring. Suction was controlled in the oedometer cell by circulating air at a constant humidity in a closed circuit including a pneumatic pump and a bottle containing a saturated salt solution, as shown in Figure 1-b.

The base plate of the oedometer cell was grooved in order to allow for vapour movements, and a thin porous stone was placed on the grooves. The sample was placed on the porous stone,

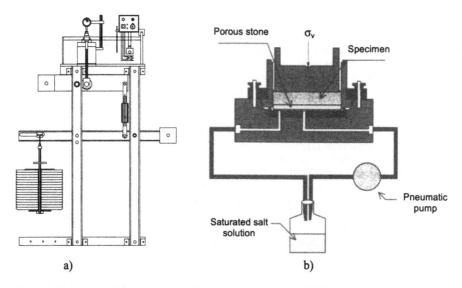

Figure 1. Device used for suction controlled oedometer tests under high stresses.
a) High pressure oedometer; b) oedometer cell with suction control using saturated salt solutions.

and vapour transfers occurred through the porous stone. In order to improve suction control and avoid any possible vapour condensation in some parts of the circuit, the cell, the bottle, and most of the ducts, excluding the pump, were immersed in a temperature controlled bath (±0.5°C). It was considered that suction equilibrium within the sample was reached when vertical displacements stabilised. This criterion is believed relevant due to the high swelling properties of the sample. The salt used were K_2CO_3, $Mg(NO_3)_2$, $NaNO_2$, $NaCl$, $ZnSO_4$ and $CuSO_4$, imposing suctions equal 113, 82, 38, 12.6 and 6.1 MPa respectively. Lower suctions were imposed using Polyethylene Glycol (PEG) solutions at various concentrations. Since PEG solutions were not saturated, their concentrations after equilibrium were measured by determining their refraction index, given in Brix degree (see Delage et al. 1998).

3 EXPERIMENTAL PROGRAM

The 3 tests performed are described by the loading paths followed in the suction versus vertical stress diagram presented in Figure 2. Samples were first incrementally loaded, at a constant suction S = 113 MPa equal to the initial suction, up to maximum vertical stresses equal to 5.2, 15.6 and 20.8 MPa respectively. After stabilisation of settlements, a wetting – drying cycle under controlled suction was applied.

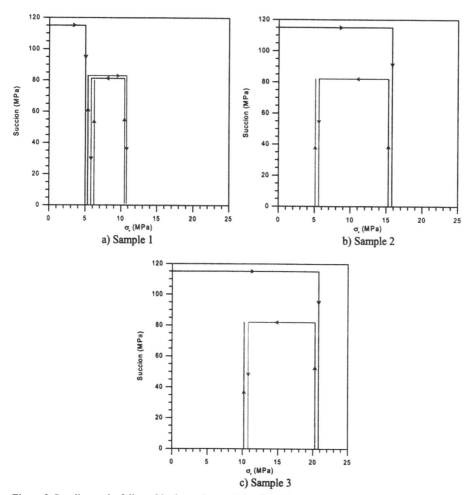

a) Sample 1

b) Sample 2

c) Sample 3

Figure 2. Loading paths followed in the various tests performed.

Under $\sigma_v = 5.2$ MPa, the suction of the first sample was decreased in steps, corresponding to suctions of 113, 82, 57, 38 and finally 0.37 MPa. The sample was subsequently dried by increasing suction in steps from 0.37 MPa to respectively 6.1, 38, 57 and 82 MPa. After stabilisation under 82 MPa was reached, the sample was loaded to $\sigma_v = 10.5$ MPa under a constant 82 MPa suction. Another suction cycle was subsequently applied under $\sigma_v = 10.5$ MPa, with suction equal to 38, 12.6, 6.1, 1.08 MPa respectively during wetting, and 6.1, 38 and 82 MPa during drying. Finally, the sample was unloaded down to $\sigma_v = 5.2$ MPa under a 82 MPa suction, and submitted to a similar suction cycle between 82 and 0.55 MPa.

The second sample, loaded under $\sigma_v = 15.6$ MPa with a suction equal to 113 MPa, was submitted to a wetting – drying cycle between 113, 0.66 and 82 MPa respectively, then unloaded down to $\sigma_v = 5.2$ MPa under a 82 MPa suction. After stabilisation of the deformation was reached, another wetting – drying cycle under a constant vertical stress ($\sigma_v = 5.2$ MPa) was applied, with suctions equal to 82, 38, 12.6, 6.1 and 0.53 MPa respectively.

The third sample was submitted to a similar suction/stress path. Initially loaded to $\sigma_v = 20.8$ MPa, the sample was submitted to a suction cycle between 113, 0.58 and 82 MPa, then unloaded down to $\sigma_v = 10.5$ MPa, and submitted to a similar suction cycle under 10.5 MPa.

The time period needed to reach deformation equilibrium during each suction step is equal to approximately 10 days, leading to a total time period of 6 months to complete each test.

4 RESULTS

Figure 3 shows the changes in void ratio obtained in the first test during the suction cycles under vertical stresses equal to 5.2 and 10.5 MPa respectively. Each step is numbered. The void ratio decrease from point 1 to 2 corresponds to the application of the 5.2 MPa load under a 113 MPa suction. The subsequent wetting path (points 2 to 7) shows the existence of three zones : the first zone (point 2 to 3) corresponds to the suction decrease from 113 to 82 MPa, it only shows a slight swelling ; the second one (points 3 to 6) corresponds to the suction decrease from 82 to 6.1 MPa, it shows a significant swelling ; the third zone (point 6 to 7) corresponds to a suction decrease from 6.1 to 0.37 MPa, and does not show any significant swelling.

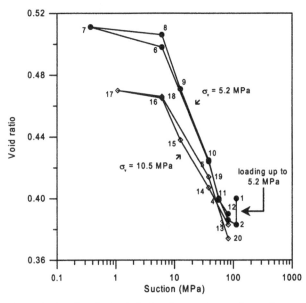

Figure 3. Results obtained during the suction cycles under constant vertical stresses ($\sigma_v = 5.2$ then 10.5 MPa).

The subsequent drying path followed under the same vertical stress (5.2 MPa, points 7 to 12) shows a good correspondence with the previous wetting curve, particularly for suctions higher than 10 MPa. Once again, a slight volume change is observed between 0.37 and 6.1 MPa. The constant suction (82 MPa) loading from 5.2 to 10.5 MPa (point 12 to 13) corresponds to a decrease in void ratio. The subsequent suction cycle (points 13 to 17 by wetting, and points 17 to

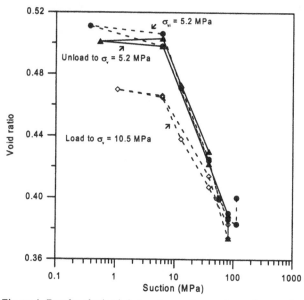

Figure 4. Results obtained during the suction cycles under constant vertical stresses (σ_v = 5.2 then 10.5 and finally 5.2 MPa).

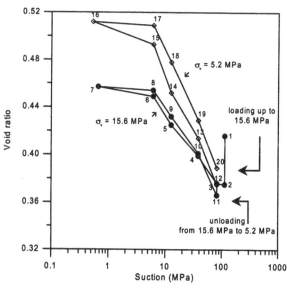

Figure 5. Results obtained during the suction cycles under constant vertical stresses (σ_v = 15.6 then 5.2 MPa).

20 by drying) also show a reversible response with small volume change below a 6.1 MPa suction, the total swelling being smaller due the higher stress applied.

Following the suction cycle under 10.5 MPa, the sample was unloaded back to 5.2 MPa under a 82 MPa suction, and a cycle in suction was imposed once again. The data obtained are compared to those of Figure 3 in Figure 4. Data are very similar, showing a reversible response, and significant swelling-shrinkage strains between 82 and 6.1 MPa. So, no influence of the cycle in suction performed under 10.5 MPa is observed, and the response of the sample submitted to suction changes under a constant load apparently does not depend on the previous suction-stress path followed.

Figure 5 shows the data obtained from the test described in Figure 2-b. The decrease in void ratio from point 1 to 2 corresponds to the application of the 15.6 MPa vertical stress, under a controlled suction of 113 MPa. As previously, the subsequent wetting path, corresponding to points 2 to 7, show again three distinct swelling zones.

The drying path (points 7 to 11) highlights a remarkable reversible volumetric deformation. Unloading from 15.6 to 5.2 MPa under a controlled 82 MPa suction (point 11 to 12) was followed by a suction cycle under σ_v = 5.2 MPa (points 12 to 20). Similar conclusions can be drawn concerning the shape of the curves, the reversibility being somewhat less evident under 5.2 MPa.

Figure 6 shows the results of the test presented in Figure 2-c. Similar conclusions can be drawn.

5 DISCUSSION

All data obtained during the first suction cycle in the previous tests (under σ_v = 5.2, 15.6 and 20.6 MPa) are gathered together in Figure 7. Reversible behaviour under suction cycles is observed under all stresses, showing an elastic behaviour under changes in suction. Smaller cyclic deformations are obtained under higher stresses, with small difference between 15.6 and 20.6 MPa. Above 6.1 MPa, the curves are linear in a semi-logarithmic plot, with a highest slope under the smallest load.

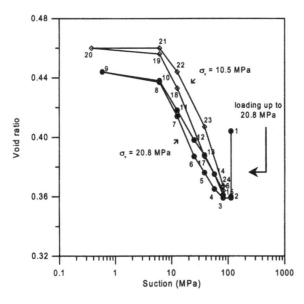

Figure 6. Results obtained during the suction cycles under constant vertical stresses (σ_v = 20.8 then 10.5 MPa).

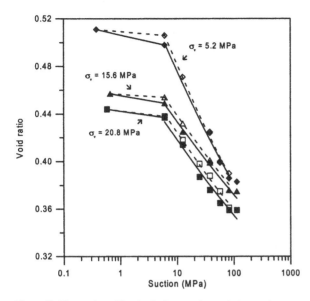

Figure 7. Change in void ratio during wetting – drying paths under various constant vertical stresses.

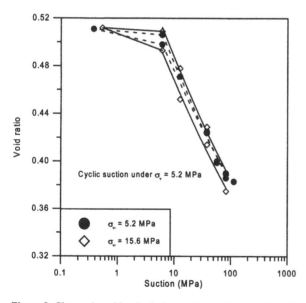

Figure 8. Change in void ratio during wetting – drying paths under constant vertical stress ($\sigma_v = 5.2$ MPa) for specimens previously submitted to two different stress paths.

Figure 8 shows the changes in void ratio under suction cycles and a similar vertical stress ($\sigma_v = 5.2$ MPa), for two specimens submitted to different previous paths. The first sample was loaded under $\sigma_v = 5.2$ MPa and submitted to a suction cycle, whereas the second sample was initially loaded under $\sigma_v = 15.6$ MPa, submitted to a suction cycle, unloaded under 5.2 MPa and submitted to a suction cycle. The figure shows a similar response, demonstrating that the volumetric deformation is not influenced by the stress path previously followed, and confirming the elastic behaviour with no stress path dependency.

71

The same conclusion can be drawn from Figure 9, which presents the changes in void ratio during a suction cycle under $\sigma_v = 10.5$ MPa for two specimens previous submitted to suction cycles under 5.2 and 15.6 MPa.

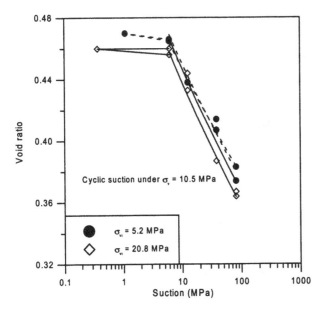

Figure 9. Change in void ratio during wetting – drying paths under constant vertical stress ($\sigma_v = 10.5$ MPa) for specimens previously submitted to two different stress paths.

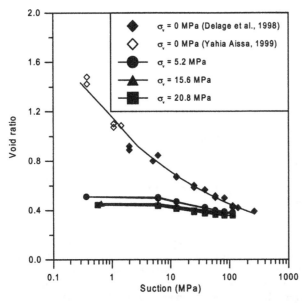

Figure 10. Evolution of the void ratios of compacted specimens under wetting – drying paths under respectively zero vertical stress and various constant vertical stresses.

The results obtained in all tests presented here show that the conclusions obtained from the study of water retention curves of FoCa7 clay (Delage et al. 1998) can also be extended to the response of samples submitted to suction cycles under a constant load. Pure swelling reversible behaviour is observed under various loads as high as 20.8 MPa, confirming the dominating influence of the physico-chemical interactions at a microscopic level, and showing no influence of the inter-aggregate porosity, generally responsible for collapse behaviour. It confirms that these macropores should not exist in the microstructure, due to the high compaction stress applied (60 MPa), and to the high density achieved (1.85 Mg/m^3). For this reason, the response obtained here differs from that obtained in other similar tests, such as those performed by Alonso et al. 1995 for instance.

A comparison between the results obtained here and the results presented in Delage et al. (1998), completed by those of Yahia-Aïssa (1999), is made in Figure 10. The figure shows that, with no stress applied, and with free deformations allowed, void ratios as high as 1.3 are reached at low suction, with no volume stabilisation at suctions lower than 6.1 MPa. Free swelling points were also obtained using PEG suction control, and the change in the technique of suction control above 6.1 MPa (saturated solutions) and below 6.1 MPa (PEG solutions) used in the tests presented here is not believed to be the reason of the difference observed in the volume change behaviour. The difference between the two series of tests (no stress applied and a constant stress applied) could be related to the one dimensional mode of deformation allowed in the tests under a constant load. This could be checked by running isotropic volume change tests under suction cycles and a constant load. However, such tests are difficult to perform, because of their long duration, and because of leakage problems which finally still occur in the confining cells.

6 CONCLUSION

The study presented here was aimed at investigating the response of samples of a dense FoCa7 compacted swelling clay submitted to suction cycles in a large range of suctions (0 – 113 MPa), under a constant load, in a high pressure oedometer. As compared to other results obtained on looser compacted swelling clays with a double structure including intra and inter aggregate porosity, no combined swelling collapse behaviour was observed here under changes in suction. This trend evidences a major influence of the microstructure level, and no effects of the inter-aggregates pores, which have been probably destroyed during the application of the high compaction stress. The remarkable reversibility of the volume changes observed under suction cycles and a constant stress is in a good agreement with the reversibility previously observed on samples free to deform under no stress. This reversibility is also related the absence of inter-aggregates pores, and to the dominant influence of physico-chemical clay-water interactions occurring at the microstructure level. Under the various constant stresses applied, a stabilisation of the deformations below 6.1 MPa was observed. More investigation seems necessary to confirm this point, and to check if it could be related to a specific hydration mechanism, in relation with the various possible adsorption mechanisms.

REFERENCES

Alonso, E. E., A. Lloret, A. Gens & D. Q. Yang 1995. Experimental behaviour of highly expansive double-structure clay. *Proceedings of the 1st International Conference on Unsaturated Soils, Paris, France*, 1: 11-16.

Atabek, R. B., B. Félix, J. C. Robinet & R. Lahlou 1991. Rheological behaviour of saturated expansive clays materials. *Workshop on Stress Partitioning in Engineered Clay Barriers, Duke University, Durham. NC.*

Belanteur, N., S. Tacherifet & M. Pakzad 1997. Etude des comportement mécanique, thermo-mécanique et hydro-mécanique des argiles gonflantes et non gonflantes fortement compactées. *Revue Française de Géotechnique* 8: 31-50.

Chen, F. H. 1975. *Foundations on expansive soils*. Elsevier Publishing Company.

Chen, X. Q., Z. W. Lu & X. F. He 1985. Moisture Movement and Deformation of Expansive Soils. *Proceedings of the 5th International Conference on Soil Mechanics and Foundation Engineering* 4: 2389-2392.

Delage, P., M. D. Howat & Y. J. Cui 1998. The relationship between suction and swelling properties in a heavily compacted unsaturated clay. *Engineering Geology* 50: 31-48.

Escario, V. & J. Sáez 1973. Measurement of the properties of swelling and collapsing soils under controlled suction. *Proceedings of the 3rd International Conference on Expansive Soils, Haifa*, 1: 195-200.

Gehling, W. Y. Y., E. E. Alonso & A. Gens 1995. Stress-path testing of expansive soils. *Proceedings of the 1st International Conference on Unsaturated Soils, Paris, France*, 1: 77-82.

Gens, A. & E.E. Alonso 1992. A framework for the behaviour of unsaturated expansive clays. *Canadian Geotechnical Journal* 29: 1013-1032.

Habib, S. A. & D. Karube 1993. Swelling Pressure behavior under controlled suction. *Geotechnical Testing Journal* 16(2): 271-275.

Pakzad, M. 1995. Modélisation du comportement hydro-mécanique des argiles gonflantes à faibles porosités. *Thèse de Doctorat à l'Université d'Orléans, France*.

Romero, E., A. Lloret & A. Gens 1995. Development of a new suction and temperature controlled oedometer cell. *Proceedings of the 1st International Conference on Unsaturated Soils, Paris, France*, 2: 553-559.

Sayouri, N. 1996. Approche microstructurale et modélisation des transfert d'eau et du gonflement dans les argiles non saturé. *Thèse de Doctorat à l'Ecole Centrale de Paris, France*.

Vilar, O. M. 1995a. Suction controlled oedometer tests on a compacted clay. *Proceedings of the 1st International Conference on Unsaturated Soils, Paris, France*, 1: 201-206.

Villar, M. V. 1995b. First results of suction controlled oedometer tests in highly expansive montmorillonite. *Proceedings of the 1st International Conference on an Unsaturated Soils, Paris, France*, 1: 207-213.

Yahia-Aïssa, M. 1999. Comportement hydromécanique d'une argile gonflante fortement compactée. *Thèse de Doctorat à l'Ecole Nationale des Ponts et Chaussées, Paris, France*.

Experimental Evidence and Theoretical Approaches in Unsaturated Soils, Tarantino & Mancuso (eds)
© *2000 Taylor & Francis, ISBN 90 5809 186 4*

Soil behaviour in the small and the large strain range under controlled suction conditions

R. Vassallo & C. Mancuso
Dipartimento di Ingegneria Geotecnica, Università degli Studi di Napoli Federico II, Italy

ABSTRACT: This paper examines the effects of suction and compaction-induced fabric on mechanical behaviour of soils. The discussion mainly focuses on: (a) the transition from saturated to unsaturated state, (b) the behaviour in the small strain range and (c) the effect of moulding water content on compacted soils' mechanical response. Laboratory tests were conducted at the Dipartimento di Ingegneria Geotecnica (DIG) of Naples on unsaturated silty sand compacted at the optimum and wet of optimum water contents. The experimental program carried out on this material included both triaxial and resonant column – torsional shear tests, all conducted under controlled suction conditions.

1 INTRODUCTION

The importance of suction in unsaturated soil in order to understand its mechanical behaviour has been widely established by the scientific community in the last decades. At present, the necessity of considering suction as a separate variable has been generally accepted. Efforts have been devoted to the understanding of general rules governing unsaturated soils' behaviour, proposing state relationships for deformation and failure problems (Fredlund 1998), as well as to the development of an elasto-plastic framework capable of fitting the main features of the experimentally observed behaviour (Alonso et al. 1990). This subject of research is relatively new and continues to develop. Work is still required in order to understand the behaviour of unsaturated soils under special conditions. For example: the transition from saturated to unsaturated state, the behaviour in the small strain range under controlled suction conditions, the effect of moulding water content on compacted soils behaviour and many others. This paper focuses on the discussion of the above mentioned areas, on the basis of quite extensive laboratory examinations on unsaturated samples prepared by compacting them to optimum and wet of optimum water contents. The experimental program included both triaxial tests and resonant column – torsional shear tests, all carried out under controlled suction conditions.

2 EXPERIMENTAL DATA AND TESTING TECHNIQUE

The tested soil is a silty sand having a clay fraction of about 16% and plasticity index PI, measured on the fraction having D_{max}<0.4mm, is around 14%. The material was compacted at optimum (w_{opt}=9.8%) and wet of optimum (w_{opt}+2.5%) water contents according to the modified Proctor procedure (ASTM D1557-91). Compaction induced a degree of saturation S_r=75% and a specific volume v=1.345 in the first case and S_r=83%, v=1.389 in the second case. Table 1 summarises the main physical and after-compaction properties of the soil.

The triaxial tests were performed by using a Bishop & Wesley cell (Rampino et al. 1999), suitable for controlling matric suction and separately measuring volumetric strains and water

Table 1. Average properties of the tested soil.

Index Properties			After Compaction at w=9.8%			After Compaction at w=12.3%		
w_L %	w_P %	PI %	v -	S_r %	u_a-u_w kPa	v -	S_r %	u_a-u_w kPa
35.4	21.7	13.7	1.345	75	800	1.389	83	60

volume changes. 23 tests were performed, in which the suction levels of 0, 100, 200, 300 kPa were considered for the optimum compacted soil and 0, 100, 200 kPa were considered for the wet compacted material. Each test started with an equalisation stage, performed at (p-u_a)=10 kPa and q=0 kPa, low enough to allow the study of the soil's response starting from compaction induced conditions. This was followed by isotropic compression and then finally by a shear stage, both conducted at constant suction value. Two types of shear stages were carried out: constant (p-u_a) and "standard" shear tests, fully drained.

Resonant-column and torsional shear tests were carried out by using a device recently developed at the DIG of Naples (Vinale 1999). The applied suction values were 0, 25, 50, 100, 200, 400 kPa for the optimum soil and 0, 100, 200, 400 kPa for the wet soil. These tests also consisted of a preliminary equalisation stage, followed by compression and shear stages. A multi-stage sequence was adopted for compression and shearing: the latter was carried out at (p-u_a) equal to 100, 200 and 400 kPa in each test. For the intermediate (p-u_a) values, torque was progressively increased not trespassing the elastic threshold strain of the soil. Once the highest scheduled (p-u_a) was achieved, torque was increased up to the maximum allowable value (i.e. 0.46 N·m) to investigate also non-linear soil behaviour.

In both devices, suction was applied using the axis translation technique. In the literature, it has often been said that this technique could not work properly if the degree of saturation is greater than about 80%, i.e. when the air in the pores may be in the form of occluded bubbles. The degree of saturation exceeded the mentioned value in some of the tests performed, i.e. at the lower applied suction values for the optimum compacted soil and in almost all the tests carried out on samples compacted to wet of optimum. Clearly, the condition S_r>80% cannot be avoided when investigating the transition from saturated to unsaturated conditions and vice versa. However, it must be highlighted that most criticisms to axis translation are based on its limitations as a technique of measuring suction, rather then controlling it. Bocking & Fredlund (1980) demonstrated, using a numerical model, that suction may be overestimated in axis translation technique since the increase in air pressure under water undrained conditions causes compression of the pore fluid and modifies the shape of the air-water interfaces. This evidently influences the pore water pressure measured at equilibrium conditions. Conversely, when suction is applied through axis translation, both air and water pressures are controlled at the specimen boundaries. In our opinion, any significant difference between the pressures of externally applied air and of the eventual occluded bubbles can be excluded, if a macroscopic equilibrium is observed at the end of the equalisation phases. In other words, it seems reasonable that after "macroscopic equalisation" the above mentioned pressures have also sensibly reached equilibrium because of air diffusion through water.

3 EFFECT OF SUCTION ON SOIL BEHAVIOUR

3.1 Basic models

In unsaturated soils the inter-granular forces are influenced by the arrangement of water and air within the voids and by the interaction between water, air and the soil skeleton. It is important to draw a distinction respect to water distribution. In a soil element there may be menisci water at the contact points between the grains and bulk water filling the voids within clusters of grains surrounded by air. In both cases pore air and water are separated by a curved interface, and therefore they are subjected to different pressures u_a and u_w, respectively. However, the way in which suction, namely the difference (u_a-u_w), affects the state of stress of the soil skeleton is very different for menisci- and bulk-water: the former influences essentially normal forces between particles while the latter has basically the same role of pore water in saturated soils. This statement can be clarified by using model of two spherical particles and considering

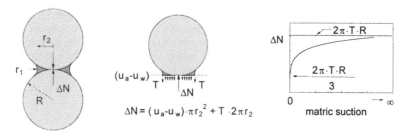

Figure 1. Water-air menisci between two solid spheres (elaborated from Fisher 1926).

a water-air meniscus between them. In Figure 1 the geometry of the meniscus is simplified, as its curvature is supposed to be constant in the plane of the drawing. From equilibrium considerations it follows that suction induces only normal forces ΔN between the particles, i.e. forces in the direction of the line connecting the spheres' centres.

Therefore, the effect of increasing suction is a greater force holding the spheres together. This has also been proved for more complex particles' configurations and meniscus' shapes (Gili 1988). Since the contacts perform frictionally, the suction-induced forces give rise to a greater slippage strength, i.e. improve the stiffness and the strength of the soil skeleton. ΔN increases as suction increases, but tends towards a threshold value. In fact, the beneficial effect of (u_a-u_w) is not indefinite because of the progressive reduction in the meniscus radius.

The model in Figure 1 also helps to understand that the effects of (u_a-u_w) and of stresses induced by external loads are uncoupled. As a matter of fact, the current $|\sigma-u_a|$ value is representative of both normal and tangential forces acting at inter-particle contacts. On the other hand, variations in suction affect only normal forces acting at grain contacts. Therefore, in unsaturated soils the Terzaghi's effective stress principle is no longer valid and a mono-tensorial approach is not possible (Bishop & Blight 1963; Fredlund & Morgernstern 1977). Consequently, two independent stress variables are required.

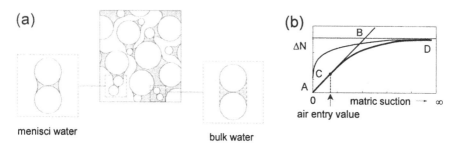

Figure 2. Effect of suction on the increment of normal force (ΔN) between two spherical particles.

Conversely, in a saturated portion of soil, such as group of particles with voids filled by bulk water, Terzaghi's principle is valid. The presence of a curved interface only implies that pore water is subjected to a pressure smaller than air pressure, i.e. suction is not zero. However, a change in suction at constant net stress varies both normal and tangential forces between particles as it represents a change in mean effective stress ($p-u_w$). In Figure 2a this situation is modelled making reference again to a couple of spheres. In the scheme air pressure u_a acts on the external half of the spheres and therefore represents the constant external stress, while water pressure u_w acts on the internal half of the spheres. As a consequence, the normal forces due to suction, ΔN, are a linear function of (u_a-u_w) equal to $(u_a-u_w) \cdot A_s$, where A_s is the cross section of the spheres.

Figure 2b shows the trend of ΔN versus suction for both bulk- and menisci- water. The behaviour of the aggregate passing from saturated to unsaturated conditions can be argued from the previous considerations, also observing that the two types of water can actually coexist in a soil having particles of different sizes, at least for a high degree of saturation. Until the air entry value s_{ev} is achieved, ΔN varies on bulk-water curve (AB), while ΔN tends toward the menisci-water curve (CD) for suction values greater than s_{ev}. The way in which a real soil moves from bulk water to menisci water depends on its response upon saturation - de-saturation.

3.2 *Observed effects on compressibility*

Compressibility has been studied through isotropic tests carried out by increasing $(p-u_a)$ at the constant rate of 4 kPa/h, sufficiently low to ensure water-drained conditions for the tested soil (Rampino et al. 1999).

Figures 3a and 3b represent the compression curves in the plane mean net stress $(p-u_a)$: specific volume v. Unloading-reloading cycles indicate that significant plastic deformations occurred during compression. Furthermore, the virgin state lines are well distinct for each suction level. The effect of suction is relevant: the compressibility decreases and the apparent pre-consolidation pressure increases as (u_a-u_w) increases. In unloading-reloading cycles the stiffening is relevant for the wet material (index k – based on natural logarithm of $(p-u_a)$ – changes from 0.0120 to 0.0067) and negligible for the optimum compacted soil (k ranges from 0.0056 to 0.0051).

Moving from saturation to 200 kPa of suction, compressibility index λ - based on natural logarithm of $(p-u_a)$ - varies from 0.022 to 0.015 for the optimum soil and from 0.040 to 0.029 for the wet soil. The values of λ are plotted in Figure 4a as a function of suction. The curves have a decreasing gradient: most changes occur in the range 0÷100 kPa while for greater suction λ tends towards a minimum threshold value: consistently with the model in Figure 1, the beneficial effect of suction is not indefinite. The experimental points are interpreted by the exponential law proposed by Alonso et al. (1990):

$$\lambda(0) = \lambda(s) \cdot \left[(1-r) \cdot e^{-\beta \cdot s} + r \right] \tag{1}$$

where s is suction and β and r determine the rate of increase of λ with (u_a-u_w) and the ratio between the threshold and the saturated value of λ, respectively. Their values for the tested silty sand are $\beta=0.024$ kPa^{-1}, r=0.68 for the optimum material and $\beta=0.015$ kPa^{-1}, r=0.73 for the wet soil. The regression lines are plotted along with experimental points in Figure 4a.

In the investigated mean stress range, all the virgin compression lines are almost linear in $\log(p-u_a)$:v plane and diverge as $(p-u_a)$ increases, as predicted by equation (1). This means that the amount of collapse phenomena is expected to be greater as mean net stress increases. Some experimental evidence indicates, on the contrary, that a maximum does exist for wetting induced collapse (Yudhbir 1982). For this reason, Josa et al. (1992) assumed that compression curves pertaining to unsaturated conditions are not straight lines in the plane $\ln(p-u_a)$:v. For the tested soil, as stated above, there is no strong indication of a change in slope in the

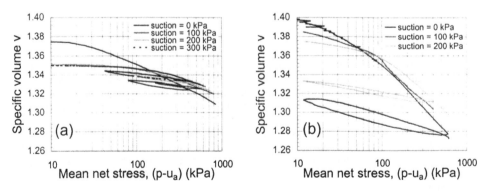

Figure 3. Isotropic compression curves: (a) optimum compacted specimens; (b) wet compacted material.

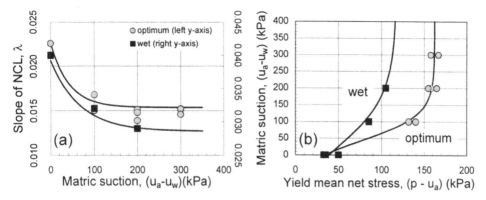

Figure 4. Effect of suction: (a) on compressibility indexes λ; (b) on apparent preconsolidation pressures.

investigated (p-u_a) range. Also oedometer tests, carried out on the same material (Rampino et al. 1999) up to a net vertical stress of 5 MPa and at suction values from 0 to 400 kPa, did not show significant variations in the slope.

The effect of suction on yield stresses is also significant. The apparent preconsolidation pressure (p-u_a)$_y$ of the optimum soil, indicated by circular data points in Figure 4b, ranges from 35 to 160 kPa when (u_a-u_w) increases from zero to 300 kPa. Again, most changes in the yield stress occur in the suction range between zero and 100 kPa and (p-u_a)$_y$ seems to tend towards a limiting value for suction greater than 200 kPa. This feature agrees with the basic model of Figure 1, indicating a decreasing gradient of suction effects. In Figure 4b data points are fitted with the regression line from the LC yield locus proposed by Alonso et al. (1990).

3.3 Shear stiffness

3.3.1 Literature overview

At present most experimental evidence about effects of suction on shear stiffness concerns the triaxial conditions and large strains. Understanding of small and medium strain behaviour of unsaturated soils is of greater importance for many engineering applications (Vinale et al. 1999). Lack of experimental evidence on this aspect is probably due to the difficulties that are encountered in developing and working with devices which actually allow to control soil suction. Therefore, data concerning the precise form of the relationship between shear stiffness and suction are rather scarce and contradictory.

Brull (1980) reports a linear relationship between initial shear stiffness G_0 and suction for a compacted silt and a compacted sand, in the range 0÷80 kPa of suction. Wu et al. (1994) performed resonant column tests on a silt without controlling suction but assessing the degree of saturation immediately after measuring stiffness. Their testing procedure consisted in applying a confining pressure on unsaturated specimen under drained conditions and measuring G_0 after 1000 minutes. Finally they extracted the specimen from the cell to measure S_r. The obtained G_0:S_r function, for a certain confining stress, shows a distinct peak, corresponding to S_r near 10÷20%. The ratio between the maximum shear modulus and the saturated value decreases as the confining pressure increases.

The experiments described above, however, did not allow to control all the stress variables affecting soil behaviour (i.e. are not performed under controlled suction conditions), so the interpretation of their results is not straightforward, as usually the observed trends of stiffness versus suction hide unknown variations of other factors. Even more complicated is the case when either water content or degree of saturation, rather than suction, is measured.

Other studies were conducted more recently under controlled suction conditions, but at null (σ-u_a). Marinho et al. (1995) performed bender elements measurements on London clay specimens assessing suction with the filter paper technique. Their results indicate a maximum in the

G_0:(u_a-u_w) relation, in the range S_r=75÷85%. Picornell and Nazarian (1998) report some results obtained on silt and clay reconstituted samples, using bender elements inside a suction plate. The Authors show that a power law can fit G_0 values versus suction and that the moduli tend to a constant value when moving towards residual water content.

Finally, Cabarkapa et al. (1999) used the bender elements technique in a triaxial cell and controlled suction via axis translation. They concluded that, for a normally consolidated quartz silt, an unsaturated G_0 value can be obtained by multiplying the saturated G_0 value pertaining to the same (p-u_a) by a factor depending only on (u_a-u_w). As a matter of fact, every G_0:(p-u_a) curve pertaining to a constant suction level is fitted by a power law with the same exponent. This implies that the ratio between two G_0 values at a certain (p-u_a) but at different suctions, such as the ratio between unsaturated and saturated values, is independent of (p-u_a) level. In other words "normalised" G_0/$G_{0,sat}$:(u_a-u_w) curves should plot in a single trend.

3.3.2 Experimental evidence in the small strain range

Figure 5 and Figure 6 show the initial stiffness G_0 versus suction, as resulting from resonant column and torsional shear tests performed on silty sand, compacted at the optimum and wet of optimum. These data seem to confirm some features of the behaviour of the soil in the large strain range (see §3.2). In fact, most of the suction effects occur in the range 0÷200 kPa of suction, and for greater (u_a-u_w) levels the stiffness seems to tend towards a threshold value.

The effect of suction is once again significant, as highlighted by G_0 values at (u_a-u_w) equal to 0 and 400 kPa: the increase in stiffness ranges from about 85% to 50% of the saturated value for the optimum material and from about 165% to 40% for the wet material depending on the net stress level. This statement is valid for both RC and TS data; the differences in stiffness values detected in TS tests (performed at a loading rate of 0.5 Hz) and RC tests are due to strain rate effects (Vinale et al. 1999).

The stiffness values of both optimum and wet compacted soils are distinct for constant stress levels and exhibit an S-shaped variation with suction. This mechanical response complies with the typical S-shape of the characteristic curve, and can be justifieded by subdividing the investigated suction range into three different zones:

I - The first zone includes low suction values, i.e. (u_a-u_w) ranging from 0 to the air entry value s_{ev} of the soil. In this zone bulk water-type effects (see §3.1) prevail. Therefore suction changes are practically equivalent to mean effective stress changes. As a consequence, the initial gradient of the G_0:(u_a-u_w) function pertinent to a certain (p-u_a) is expected to be roughly equal to that of the G_0:p' curve of the the saturated material at p' equal to that particular (p-u_a) value. According to literature evidence, the G_0 versus p' relationship has the shape displayed in Figure 7 (d'Onofrio et al. 1999), where the gradients at two values of mean effective stress are highlighted. It is evident that the gradient is a decreasing function of mean stress.

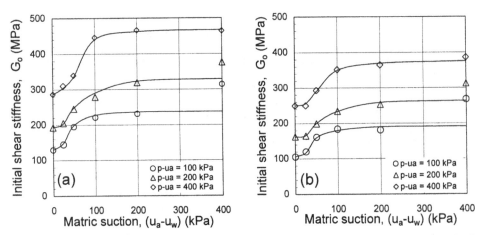

Figure 5. Initial shear stiffness against matric suction in suction controlled RC (a) and TS (b) tests (optimum compacted material).

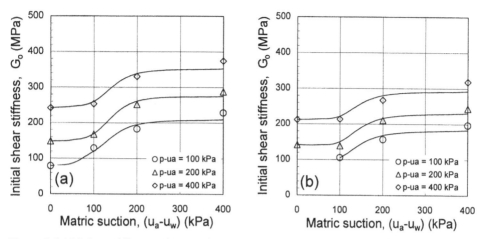

Figure 6. Initial shear stiffness against matric suction in suction controlled RC (a) and TS (b) tests (wet compacted material).

II - In the second zone, i.e. intermediate suction values, air starts to enter the pore voids in significant amount. Increasing suction determinates a progressive shift from bulk water- to menisci water- regulated soil behaviour.

III - The third zone is the region where suction is so high as to make menisci water effects (see §3.1) prevailing. The effect of (u_a-u_w) on soil behaviour agrees with the predictions of the simple model in §3.1. In the investigated suction range, the relationship between initial shear stiffness and suction has a decreasing gradient and seems to approach a limit value as (u_a-u_w) increases. Furthermore, in this zone the observed response complies with other recent experimental results (Marinho et al. 1995; Picornell & Nazarian 1998), mentioned above. In the investigated suction range (0÷400 kPa) the degree of saturation always exceeds 74%, and G_0 always increases with suction. However, this does not exclude the existence of a maximum in the G_0:(u_a-u_w) curves for lower S_r values and if soil stiffness should be predicted at very high suction more experimental evidence would be necessary.

Figure 8 shows the initial stiffness of the wet material, resulting from RC tests, normalised with respect to the relevant saturated G_0 values. This diagram shows that the S-shaped curves do not reduce to a unique trend after normalisation. As a matter of fact, while at $(p-u_a)$=100

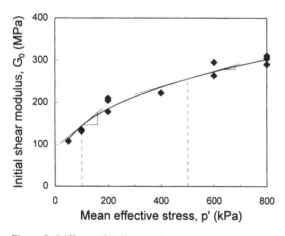

Figure 7. Stiffness of Vallericca clay versus mean effective stress (elaborated from d'Onofrio et al.1999).

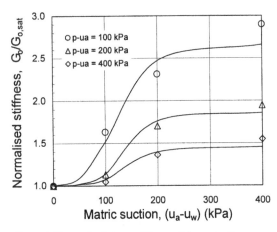

Figure 8. Normalised shear stiffness of the wet soil versus suction (RC tests).

kPa G_0 increases of 165% as suction varies from 0 to 400 kPa, at $(p-u_a)$=400 kPa this effect reduces to 40%. It can be argued that the effect on G_0 of any fixed increment of suction decreases as $(p-u_a)$ increases. This is also valid for the optimum compacted soil, although the differences between the normalised curves are less evident. Furthermore, this seems to agree with the results of Wu et al. (1994), where they suggested that the ratio between the maximum and the saturated value of G_0 decreases as $(p-u_a)$ increases. However, it must be noted that in their tests no attempt was made to measure or control suction.

The non-linear and non-reversible pre-failure behaviour of the examined soil, derived from RCTS tests, was studied in detail by Vinale et al. (1999), who analysed the effects of suction on the curves: equivalent shear stiffness G versus shear strain γ and damping ratio D versus shear strain γ. Here it is worth noting that the improvement of soil stiffness due to suction persists in all the investigated strain range ($\gamma = 7 \cdot 10^{-4} \div 3 \cdot 10^{-2}$ %). At constant suction level, G is roughly constant and attains its maximum value (G_0) up to a "linear threshold strain" γ_L.

Contrary to expectations from the basic model of §3.1, the normalised curves $G/G_0(\gamma)$ have essentially the same shape when curves at different suction level and at $(p-u_a)$=400 kPa are compared, as done by Vinale et al. (1999).

3.3.3 Modelling small strain behaviour
The saturated values of G_0 for the tested silty sand can be modelled the equation first proposed by Hardin (1978):

$$\frac{G_0}{p_a} = S \cdot \left(\frac{p'}{p_a} \right)^n \cdot f(e) \qquad (2)$$

where p_a is the atmospheric pressure, p' is the mean effective stress, and $f(e)$ is a scaling function for void ratio-induced heterogeneity. The parameters S and n represent the stiffness of the material under the reference pressure and the sensitivity of the stiffness to the stress state, respectively (Hardin 1978). When $f(e)=1$ is assumed [the observed changes in void ratio of the tested soil are very limited (Vinale et al. 1999)], RC data yield S=1298 and n=0.57.

If the normalised shape of the G_0:suction relationship were unique, as resulting from the data of Cabarkapa et al. (1999), it would be possible to extend equation (2) to the unsaturated soil case by simply assuming S as suction dependent:

$$\frac{G_0}{p_a} = S(u_a - u_w) \cdot \left(\frac{p - u_a}{p_a} \right)^n \cdot f(e) \qquad (3)$$

and f(e)=1. The above relationship does not agree with the experimental data collected on silty sand (see fig. 8). Therefore, an alternative formulation is proposed.

In agreement with the previous discussion about the physical meaning of the S-shaped G_0:(u_a-u_w) relation, it follows that the null suction stiffness $(G_0)_{s=0}$ is obtained from equation (2):

$$\frac{(G_0)_{s=0}}{p_a} = S \cdot \left[\frac{(p-u_a)_C}{p_a} \right]^n \cdot f(e) \tag{4}$$

where $(p-u_a)_C$ is the mean net stress at which the soil has been consolidated prior to shearing and s is matric suction.

Equation (4) should not be valid only at s equal to zero, but should also hold up to the air entry value of the tested soil. Therefore, in Zone I, the G_0 versus suction relationship should follow the equation:

$$\frac{(G_0)_{s \leq s_{ev}}}{p_a} = S \cdot \left[\frac{(p-u_a)_C + s}{p_a} \right]^n \cdot f(e) \tag{5}$$

Further increases in suction cause progressive soil de-saturation and move the G_0:(u_a-u_w) relationship from the bulk- to the menisci-water regulated zone, i.e. give rise to a "zone II behaviour". In this zone, it can be reasonably assumed that the effect of suction starts with a gradient equal to that defined by equation (4), that is:

$$\left[\frac{\partial(G_0)}{\partial s} \right]_{s_{ev}} = \frac{\partial}{\partial p'} \left\{ S \cdot \left[\frac{p'}{p_a} \right]^n \cdot f(e) \right\}_{p_{ref}} = S \cdot n \cdot \left[\frac{p_{ref}}{p_a} \right]^{n-1} \cdot f(e) \tag{6}$$

with $p_{ref} = (p-u_a)_c + s_{ev}$. The G_0:s curve must link zone I and zone III functions. However, both its shape and the (u_a-u_w) value at which menisci-water effects start to prevail (i.e. the edge of zone II) depend on the details of the specific de-saturation process and are not straightforward.

Thus, a simplified formulation is proposed in this context. It is assumed that on every curve it is possible to identify one point, at a suction level called s*, characterising the transition from bulk to menisci water regulated behaviour. This point, for instance, could be the point of inflection of the S-shaped curve (as it will be clarified in the following, its precise position is not so important). On the basis of the above discussion, it can be said that the s* value should fall into zone II. For (u_a-u_w)<s*, initial shear stiffness can be obtained from equation (5). On the other hand, for (u_a-u_w)>s*, experimental points can be fitted by an exponential law, such as:

$$G_0 = (G_0)_{s^*} \cdot \left\{ [1-r] \cdot e^{-\beta \cdot (s-s^*)} + r \right\} \tag{7}$$

where β is a parameter that controls the rate of increase of soil stiffness with suction and r defines the ratio between shear stiffness G_0 at (u_a-u_w)=s* and the threshold value for increasing suction.

Equation (7) is analogous to equation (1), proposed by Alonso et al. (1990) to interpret the variation of compressibility index λ with suction for the tested material. However, it does not follow directly from equation (1). In fact, an analogy should be drawn between shear stiffness and bulk modulus K. Thus, G_0 should vary inversely with λ function at any constant stress level as K does. However, the shape of the function obtained this way would not be very different from that deriving from equation (7). Therefore the latter is preferred, since it is more manageable and yields straightforward r and β coefficients.

Although the above simplifications imply a discontinuity in the G_0:(u_a-u_w) curve at (u_a-u_w)=s*, they allow to describe with sufficient approximation the experimental results. As an example, Figure 9 displays the data obtained from RC tests performed on the optimum and the wet compacted materials at ($p-u_a$)=400 kPa, along with lines obtained using the proposed interpretation criteria. The figure shows that equation (7) can properly fit experimental data starting from stiffness values higher than those predicted by equation (5) extended to zone II.

If the observed behaviour is confirmed by further experimental evidence, one may conclude

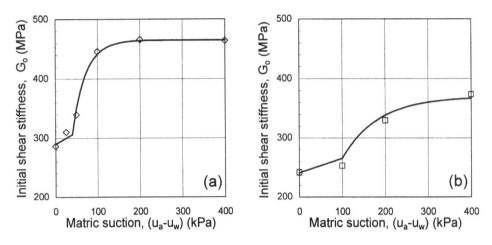

Figure 9. Interpretation of the results of the RC tests carried out on the optimum compacted material (a) and on the wet material (b) at mean net stress 400 kPa.

that menisci water effects dominate those of bulk water at relatively high suction, while low suction levels (i.e. relatively high degree of saturation) lead to the contrary. In the former case, suction increases the normal forces acting on the particles, while in the latter case suction is roughly equivalent to a change in mean net stress, which has a smaller effect on shear stiffness (Vinale et al. 1999).

3.3.4 *Experimental evidence on large strain behaviour*
Stiffness of the tested silty sand can be also obtained from data pertaining the large strain range. For instance, Figure 10 shows three $q:\varepsilon_a$ curves obtained from standard drained triaxial tests, performed on samples compacted optimum and tested at suction values of 0, 100, 200 and 300 kPa and $(p-u_a)=400$ kPa. This figure is representative of the soil response observed in all the tests. The beneficial effect of suction on stiffness is once again evident.

Secant moduli can be calculated from triaxial data at a selected level of axial strain. It is obvious that no direct comparison can be made with G_0 moduli described above. In fact, standard triaxial shear tests and RCTS tests are carried out following different stress paths. Furthermore, accurate values of stiffness can be obtained from triaxial data only for $\varepsilon \geq 0.1\%$ (Rampino et al. 1999), that is one order of magnitude greater then the maximum γ of RCTS tests. Clearly, at

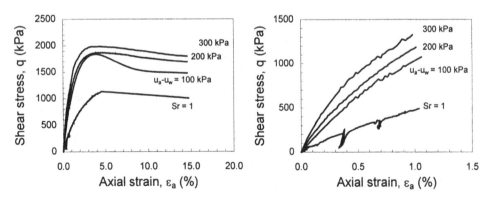

Figure 10. Deviator stress: axial strain curves from shear tests performed on the optimum compacted material at $(p-u_a) = 400$ kPa.

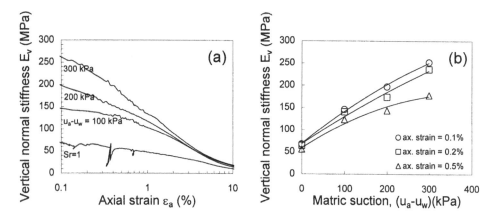

Figure 11. Vertical normal stiffness from triaxial tests performed on the optimum compacted soil: (a) versus axial strain; (b) versus suction.

such different strain levels shearing involves different mechanisms of deformation at the particle's scale. In particular, the tests of Figure 10 were performed on normally consolidated samples and thus, in this case, shearing induces macroscopic soil yielding.

The secant moduli E_V, calculated from the data in Figure 10, are plotted in Figure 11a as a function of axial strain and in Figure 11b as a function of suction, at three different strain levels. These diagrams highlight that elasto-plastic moduli at (u_a-u_w)=300 kPa are more than 3 times greater than the saturated value, i.e. indicate a more pronounced suction effect on soil stiffness, if compared to RCTS results. Furthermore, Figure 11 shows that suction influence on stiffness is not circumscribed to the range 0÷300 kPa but is probably significant also for greater (u_a-u_w) levels.

4 EFFECT OF COMPACTION INDUCED FABRIC ON SOIL BEHAVIOUR

Thus far, the optimum and wet of optimum soils have been presented as different materials, even though the mineralogical and physical characteristics of their grains are the same. In fact, the two soils are significantly different, as highlighted by the observed mechanical response. This can be explained as the influence of moulding water content on compaction induced fabric.

In literature, this effect has been studied on the basis of direct observations through S.E.M. or mercury intrusion porosimetry measurements (Delage et al. 1996) which indicate that, depending on compaction water content, fine grained soils may assume a different fabric, due to water-soil particles interaction at a microscopic scale. Dry compaction generally results in a soil fabric made up of aggregates of varying size, and usually with a bimodal pore size distribution. Differently, wet of optimum compaction tends to produce a more homogeneous matrix-dominated soil fabric and a single pore size distribution.

Besides affecting fabric, the selected preparation procedure influences the soil's initial suction, i.e. its initial stress state. To isolate the effects of fabric, it is necessary to investigate the mechanical response of differently compacted soils under the same conditions of suction and testing path. Therefore, suction control is required. At present, this kind of experimental evidence is very limited (Gens et al. 1995), probably due to the need for more sophisticated experimental apparatuses designed for unsaturated soil testing.

4.1 Isotropic stress paths

A first effect of moulding procedure is the opposite behaviour exhibited by the two materials in the equalisation stages (Mancuso et al. 2000), even though this can be mainly attributed to

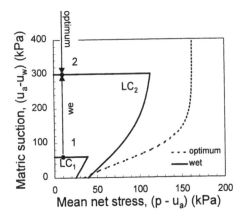

Figure 12. Evolution of the LC yield loci of the optimum and the wet soil during equalisation.

the different after compaction suctions. Using the Imperial College tensiometer, suction was measured to be 800 kPa for the optimum material and 60 kPa for the wet specimens. These values justify the observed response: the optimum soil swells and absorbs water as it follows a wetting path, while the wet material shrinks and expels water as it follows a drying path.

The influence of fabric on basic soil properties becomes evident when isotropic compression data are considered. Compressibility indices $\lambda(s)$ and $\kappa(s)$ almost double going from the optimum to the wet of optimum material, as indicated in Figures 3 and 4a. Therefore, wet compaction seems to induce a weaker soil fabric, as will be also confirmed by the following observations.

The yield points detected from the compression curves have been shown in Figure 4b. At the applied suction levels the yield net stress $(p-u_a)_y$ is always greater for the optimum soil, except for $S_r=1$, where $(p-u_a)_y$ is sensibly the same for the two materials. In fact, the yield stress ranges from 35 to 160 kPa for the optimum soil and from 39 to 105 kPa for the wet soil.

The effect of fabric on the yield stresses could be even greater than displayed in Figure 4b. In fact, each of this stresses was obtained from isotropic compression results and so is representative of an after-equalisation state (Sivakumar & Doran 2000). The optimum soil swells and reduces its suction during equalisation, thus, in this stage, its deformations can be considered reversible. Therefore, the after-compaction yield locus is the same line fitting the detected yield points in Figure 4b, plotted again in Figure 12 as a broken line. Oppositely, during equalisation the wet soil shrinks and increases its suction, as indicated in Figure 12 for equalisation at 300 kPa of suction. This process does not induce yielding only under the hypothesis that LC locus and SI locus movements are independent from one another (Alonso et al. 1990). On the other hand, if the LC and SI movements were coupled, equalisation would provoke an expansion of the elastic region of the wet soil (i.e. a shift from LC_1 – the after-compaction yield locus – to LC_2) dependent on the imposed suction value. This would imply, as written above, a difference between the LC loci of the two soils greater than detected from compression stages.

As a consequence of the considerations made above, it can be said that in its after compaction state the wet soil may be more sensitive to collapse on wetting then the optimum soil. In other words if the two materials are subjected to the same wetting path, the wet compacted soil will experience collapse starting from greater suction levels, as a consequence its smaller pseudo-elastic region.

4.2 Shearing

Other significant indications on the effect of moulding water content come from the results collected in the small strain range. As an example, Figure 13 shows the G_0:(u_a-u_w) curves obtained for both materials, at $(p-u_a)=200$ kPa, in RC tests. First of all, an increase in compaction water content causes a reduction in G_0. The initial stiffness of saturated samples, displayed in

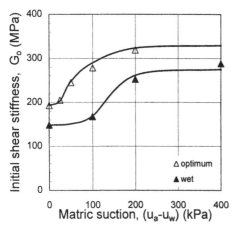

Figure 13. Initial shear stiffness against matric suction in RC tests preformed at $(p-u_a)$=200 kPa.

Figure 13, varies from about 190 MPa in the optimum case to 140 MPa for the wet compacted soil, that is a decrease of about 25%. Similar effects are observed for the other suction levels.

Furthermore, as also resulting from Figures 5 and 6, moulding water content affects the shape of G_0:(u_a-u_w) curves, i.e. the ratio between the saturated value and the unsaturated threshold values of shear stiffness and the suction value (s*) characterising transition between bulk- and menisci-water regulated behaviour (see §3.3.3). In particular, the s* level seems to be near 30÷40 kPa for the optimum material and 90÷100 kPa for the wet soil. This latter point is consistent with the results presented by Vanapalli et al. (1996) regarding the characteristic curves of a glacial till (a sandy clay) compacted at different water contents. As a matter of fact, compaction at optimum water content leads to a lower air-entry value than compaction at wet of optimum, an effect attributable to the optimum soil's more open macropore structure.

Vinale et al. (1999) indicate that the influence of fabric on non-linear behaviour is less evident: the linear threshold strain of the optimum compacted material ($\gamma_L \cong 1 \cdot 10^{-3}$%) is lower than the wet compacted value ($\gamma_L \cong 2 \cdot 10^{-3}$%), but this difference is quite small and cannot confidently confirm any effect of preparation water content on compacted soil linear response.

Mancuso et al. (2000) report that the weakening effect of a greater moulding water content on the tested soil is significant also in triaxial shear stages. As fewer data are available on the wet soil and, in addition, they are not all homogeneous due to differences in the over-consolidation ratios, this effect can be better represented by considering ultimate state points instead of stiffness. In fact, if the equation proposed by Wheeler & Sivakumar (1995):

$$q = M(p - u_a) + \mu(s) \tag{8}$$

is used to fit ultimate q and $(p-u_a)$ values, the reduction of the soil's strength due to fabric can be interpreted as a decrease in the apparent cohesion $\mu(s)$ at constant M coefficient. Cohesion varies from 0 to 240 kPa for the optimum soil and from 0 to 150 kPa for the wet soil, in the investigated suction range. This circumstance allows to strengthen some previous observations: according to most known elasto-plastic models (Alonso et al. 1990; Wheeler and Sivakumar 1995; Cui & Delage 1996) the reduction of $\mu(s)$ implies a reduction in the size of the yield locus. Therefore, it can be concluded that in p:q:s space the after-compaction pseudo-elastic region is more extended for the optimum compacted soil.

5 CONCLUSIONS

A wide experimental program carried out on a silty sand compacted at the optimum and wet of optimum water contents has been the starting point of a discussion about several aspects of soil behaviour. The transition from saturated to unsaturated state, the behaviour in the small strain range under controlled suction conditions, the effect of moulding water content have been analysed on the basis of the results obtained in triaxial and RCTS tests.

The stiffening effect of suction was justified by making reference to the basic model of two spherical particles. Suction significantly affects soil behaviour both in the small and in the large strain range. With regard to the former, the measurement of initial shear stiffness G_O in the range $0\div400$ kPa of suction allowed to point out that menisci water effects dominate those of bulk water at relatively high suction, while low suction levels lead to the contrary. As a matter of fact, the $G_O{:}(u_a\text{-}u_w)$ relation is S-shaped in accordance with the typical S-shape of the water retention curve. A simplified formulation has been proposed to fit this behaviour.

In the large strain range, it was shown that suction severely affects both compressibility and shear stiffness. The variations of compressibility index λ mostly occur in the range $0\div100$ kPa of suction, while for greater suction λ tends towards a threshold value. The beneficial effect of suction on elasto-plastic secant moduli E_v resulted to be more pronounced than the effect observed in the small strain range.

The differences in the behaviour of the optimum and the wet materials have been explained as the influence of moulding water content on soil fabric. Wet compaction seems to induce a weaker fabric, as indicated by higher compressibility and lower stiffness. Furthermore, in its after compaction state the wet soil has a smaller pseudo-elastic region. Compaction procedure influences not only G_O values but also the shape of $G_O{:}(u_a\text{-}u_w)$ curves. In fact, both the ratio between the saturated and the unsaturated threshold value of G_O and the suction value characterising the transition between bulk- and menisci- water regulated behaviour are significantly affected by moulding water content.

In summary, the results obtained on the tested silty sand have permitted to clarify some features of unsaturated soils behaviour that have been not much investigated thus far. Particular emphasis has been devoted to the role of bulk- and menisci- water effects in the transition from saturated to unsaturated conditions. This topic is relatively new and, probably, further experimental data are required to confirm the interpretation given in this context and to propose a comprehensive model.

REFERENCES

Alonso, E.E., A. Gens & A. Josa 1990. A constitutive model for partially saturated soils. *Geotechnique* 40(3): 405-430.

ASTM D1557-91. *Test method for laboratory compaction of soils using modified effort, 5600 ft·lbf/ft³ (2700 kN·m/m³), Philadelphia.*

Bishop, A.W. & G.E. Blight 1963. Some aspects of effective stress in saturated and partially saturated soils. *Geotechnique* 13(3): 177-197.

Bocking, K.A. & D.G. Fredlund 1980. Limitations of the axis translation technique. *IV International Conference on Expansive Soils, Denver* 1: 117-135.

Brull, A. 1980. Caracteristiques mécaniques des sols de fondation de chaussées en fonction de leur état d'humidité et de compacité. *International Conference on Compaction, Paris* 1: 113-118.

Cabarkapa, Z., T. Cuccovillo & M. Gunn 1999. Some aspects of the pre-failure behaviour of unsaturated soil. *II International Conference on pre-failure behaviour of geomaterials, Turin* 1: 159-165.

Cui, Y.J. & P. Delage 1996. Yielding and plastic behaviour of an unsaturated compacted silt. *Geotechnique* 46(2): 291-311.

D'Onofrio, A., F. Silvestri & F. Vinale 1999. Strain rate dependent behaviour of a natural stiff clay. *Soils and Foundations* 39(2): 69-82.

Delage, P., M. Audiguier Y.D. Cui & M.D. Howat 1996. Microstructure of a compacted silt. *Canadian Geotechnical Journal* 33: 150-158.

Fisher R.A. 1926. On the capillary forces in an ideal soil, *Journal of Agricultural Science* 16: 492-505.

Fredlund D.G., 1998. Bringing unsaturated soil mechanics into engineering practice. *II International Conference on Unsaturated Soils, UNSAT '98, Beijing* 2: 1-36.

Fredlund, D.G. & N.R. Morgernstern 1977. Shear state variables for unsaturated soils. *Journal of Geotechnical Engineering Division, ASCE* 103(5): 447-466.

Gens, A., E.E. Alonso, J. Suriol & A. Lloret 1995. Effect of structure on the volumetric behaviour of a compacted soil. *I International Conference on unsaturated soils, UNSAT '95, Paris* 2: 83-88.

Gili, Y.Y. 1988 – Modelo microestructural para medios granulares ni saturados (in Spanish). *Doctoral Thesis*, Universitat Politècnica de Catalunya.

Hardin, B.O. 1978. The nature of stress-strain behaviour for soils. State of the Art, Geotechnical Engineering Division. *Specialty Conference on Earthquake Engineering and Soil Dynamics, ASCE*, Pasadena, California.

Josa, A., A. Balmaceda, A. Gens & E.E. Alonso 1992. An elasto-plastic model for partially saturated soils exhibiting a maximum of collapse. *III International Conference on Computational Plasticity, Barcelona* 1: 815-826.

Mancuso, C., R. Vassallo & F. Vinale 2000. Effects of moulding water content on the behaviour of an unsaturated silty sand. *Asian Conference on Unsaturated Soils*, Singapore.

Marinho, E.A.M., R.J. Chandler & M.S. Crilly 1995. Stiffness measurements on an high plasticity clay using bender elements. *I International Conference on Unsaturated Soils, UNSAT '95, Paris* 1: 535-539.

Picornell, M. & S. Nazarian 1998. Effects of soil suction on the low-strain shear modulus of soils. *II International Conference on Unsaturated Soils, UNSAT ' 98, Beijing* 2: 102-107.

Rampino, C., C. Mancuso & F. Vinale 1999. Mechanical behaviour of an unsaturated dynamically compacted silty sand. *Italian Geotechnical Journal* 33(2): 26-39.

Sivakumar,V. & P.G. Doran 2000. Yielding characteristics of compacted clay. *Mechanics of cohesive-frictional materials* 5(4): 291-303.

Vanapalli, S.K., D.G. Fredlund & D.E. Pufahl 1996. The relationship between the soil-water characteristic curve and the unsaturated shear strength of a compacted glacial till. *Geotechnical Testing Journal* 19 (3): 259-268.

Vinale, F., A. d'Onofrio, C. Mancuso, F. Santucci De Magistris & F. Tatsuoka 1999. The prefailure behaviour of soils as construction materials. *II International Conference on pre-failure behaviour of geomaterials*, Turin.

Wheeler, S.J. & V. Sivakumar 1995. An elasto-plastic critical state framework for unsaturated soil. *Geotechnique* 45 (1): 35-53.

Wu, S., D.H. Gary & F.E. Richart 1989. Capillary effects on dynamic modulus of sands and silts. *Journal of Geotechnical Engineering Division, ASCE* 110(9): 1188-1203.

Yudhbir, B. 1982. Collapsing behaviour of residual soils. *VII Southest Asian Geotechnical Conference, Hong Kong* 1: 915-930.

Experimental Evidence and Theoretical Approaches in Unsaturated Soils, Tarantino & Mancuso (eds)
© *2000 Taylor & Francis, ISBN 90 5809 186 4*

Retention curves of deformable clays

E. Romero & J. Vaunat
Departamento de Ingeniería del Terreno, Cartográfica y Geofísica, Universidad Politécnica de Cataluña, Barcelona, Spain

ABSTRACT: Main wetting and drying paths in terms of water ratio (volume of water to volume of solids) indicate a delimiting zone in the water retention curve of an aggregated clayey fabric separating a region of intra-aggregate porosity from an inter-aggregate porosity adjoining area. In the intra-aggregate region, water ratio is not dependent on void ratio and retention curve parameters are mainly dependent on specific surface. However, in the inter-aggregate region, water ratio is dependent on void ratio and strongly coupled to mechanical actions. The paper presents water retention curves in the inter-aggregate region under varying void ratios of expansive and collapsible clay fabrics. These results are described and interpreted in a multiple diagram, which includes the following plots: retention curve plane, isotropic compression plane, stress path plane and volumetric state variable plane. Reversible and irreversible features of water retention characteristics are also presented and discussed.

1 INTRODUCTION

The conceptual understanding of the water retention curve is an essential element for the comprehension of unsaturated soil behavior, as well as for tackling with an increasingly wider range of geotechnical engineering applications. The water retention curve expressed in terms of water content or degree of saturation is stress-path dependent and displays an important hysteresis response. An application has recently been described by Ng & Pang (2000), who analyzed the influence of stress states on retention curves and their consequences on slope stability. However, the role of hydraulic hysteresis in the retention curve, as well as its influence on the mechanical behavior is a topic that has not been systematically taken into account yet in most existing constitutive models. This hydraulic hysteresis has a notable influence on the mechanical behavior, because of the different arrangements of water within the voids that affect the mechanical behavior of soil skeleton in different ways (Wheeler & Karube 1996).

There are a number of laboratory results and conceptual models concerning retention curves and recent works on this area are described in Leong & Rahardjo (1997), Aubertin et al. (1998), Barbour (1998), Vanapalli et al. (1999) and Dineen & Ridley (1999). In contrast, experimental information concerning the influence of mechanical actions on water retention in unsaturated soils is limited. Swelling-collapse, shrinkage and loading paths on soils, as well as contracting or dilating behavior as a consequence of shear strains, affect their water storage capacity. Recent experimental results showing this hydro-mechanical coupling are presented in Wheeler (1996) with reference to isotropic loading tests, in Rampino et al. (1999) with relation to triaxial compression and deviatoric stages, Chen et al. (1999) concerning isotropic compression and shrinkage tests, and Ng & Pang (2000) with relation to drying and wetting paths in oedometer cells. Kawai et al. (2000) have also presented retention curve results of a silty clay considering effects of void ratio changes. Biarez et al. (1988) and Fleureau et al. (1993) have also reported results showing the volume change behavior and water content evolution of clayey soils in drying-wetting paths. In addition, Romero (1999) has presented wetting and drying paths, as well as loading and unloading paths in oedometer and isotropic cells, showing coupled consequences on volumetric behavior and water content changes in unsaturated clays. Therefore, it is necessary to improve current equipment and experimental techniques, as well as constitutive models, to investigate coupled hydro-mechanical aspects under unsaturated states and to provide a reliable description of the stress-strain behaviour of the soil. Two aspects are required to be considered in detail for isotropic stress states: firstly, the response of the retention curve during wetting and drying cycles at constant net mean stresses, and secondly, upon loading and unloading paths at a constant matric suction.

The paper presents constant volume main wetting and drying retention curves in terms of water ratio (volume of water to volume of solids), which have been obtained for different clay fabrics at two constant void ratios, statically compacted at dry of optimum and using vapor equilibrium and air overpressure techniques. This latter method translates water pressure in the positive range, the same as axis translation technique, but regulates suction under a constant air pressure and variable water pressures. Experimental data have been obtained on aggregated fabrics under the assumption that stress paths have their major influence on retention curves for this type of structure compared to matrix dominant fabrics. Retention curves under varying void ratios in expansive and collapsible clay fabrics are also presented. Here information is complemented with stress paths and volumetric state variable evolutions. In addition, experimental results showing irreversible aspects on water ratio under isotropic loading and unloading paths, as well as on wetting and drying paths, are also described. Test results are examined and main features of behavior discussed. Experimental data describing the variation of water ratio under suction and net stress paths will be the background of a constitutive model, which will be described in a companion paper (Vaunat et al. 2000, this issue).

2 DEFINITION OF STRESS AND WORK CONJUGATE STRAIN VARIABLES

The approach followed in this paper is to use two independent stress state variables to describe the behavior of isothermal, unsaturated and chemically inert clays: the net stress state variable $(\sigma_{ij}-u_a\delta_{ij})$ and the matric suction stress state variable $(u_a-u_w)\delta_{ij}$, where the first tensor is the total stress, and u_a and u_w are the gas and liquid phase pressures. This combination results advantageous because under a constant air pressure (air overpressure technique), which is the case of the experimental program, the effects of a change in net stress, inducing changes on contact forces between aggregates, can be separated from the effects caused by pore-water pressure affecting soil skeleton via interface actions. For axially symmetric conditions, three independent stress parameters are defined: the net mean stress $[(\sigma_v+2\sigma_h)/3-u_a] = (\sigma_m-u_a)$, the deviatoric stress $(\sigma_v-\sigma_h)$ and the matric suction $s = (u_a-u_w)$, where σ_v and σ_h are the principal axial and radial total stresses, respectively.

Assuming an isotropic stress state σ_m, the work input to the soil per unit of initial volume δW by means of the application of total stress, water and air pressures, can be expressed in a simple way as (solid and water phases are regarded as incompressible):

$$\delta W = \sigma_m \delta\varepsilon_v + u_w\delta\theta_w + u_a\delta\theta_a + u_a\theta_a \frac{\delta\rho_a}{\rho_a} \qquad (1)$$

where positive compression doing positive work results in a volume decrease and ρ_a is the density of the air phase. $\theta_w = nS_r$ and $\theta_a = n(1-S_r)$ represent the volumetric water and air contents, where n = porosity and S_r = degree of saturation. Equation 1 is equivalent to that proposed by Houlsby (1997), neglecting the work dissipated by the flow of water and air through the soil. The positive work input to compress the air phase is given by $u_a\theta_a\delta\rho_a/\rho_a$. It appears that the first three terms of Equation 1 are sufficiently adequate to describe work input per unit volume providing that air overpressure technique is carried out under $\delta u_a = 0$ conditions. Assuming incompressibility of the solid phase (volumetric strain is only due to the variation of porosity in a Lagrangian sense as defined by Kaczmarek & Hueckel 1998: $\delta\varepsilon_v = -\delta n$), Equation 1 may be rewritten as (work input to compress the air phase is not considered):

$$\delta W = \left(\sigma_m - u_a\right)\delta\varepsilon_v - \left(u_a - u_w\right)\delta\theta_w \quad or \tag{2}$$

$$(1+e_o)\,\delta W = -\left(\sigma_m - u_a\right)\delta e - \left(u_a - u_w\right)\delta e_w \tag{3}$$

Equation 2 indicates the appropriate work conjugate extensive variables associated with different significant intensive stress variables. Matric suction is work conjugated with the extensive strain variable $-\delta\theta_w = -n\,\delta Sr + Sr\,\delta\varepsilon_v$ when incorporating the net mean stress (Edgar 1993, Wheeler and Sivakumar 1995, Dangla et al. 1997, Houlsby 1997). In this paper and according to Equation 3, the water ratio (volume of water to volume of solids) $-\delta e_w = -G_s\,\delta w$ has been selected as the volumetric state variable associated with matric suction, in the same way as the volumetric state variable void ratio $-\delta e$ is associated with the net stress variable. Plots in the conventional stress path $(u_a\text{-}u_w):(\sigma_m\text{-}u_a)$ and isotropic compression $e:(\sigma_m\text{-}u_a)$ planes are complemented with graphs in the retention curve $(u_a\text{-}u_w):e_w$ and the volumetric state variable $e:e_w$ planes. The advantage of using the volumetric state variable e_w is that it tends to the void ratio e under nearly saturated states $(S_r = e_w / e \to 1)$. e_w is estimated directly in laboratory measuring the inflow or outflow of water into a soil with a known volume of solids. In addition, e_w remains constant in water undrained tests in an equivalent concept to isochoric tests, where e remains constant.

The volumetric state variable e_w has been used in different contexts by different authors working in unsaturated soils. Toll (1995), describing a conceptual model for drying and wetting paths, used e_w to represent the gravimetric water content of an unsaturated soil on the same plot as void ratio. Prashanth et al. (1998) called it water void ratio or water volume content. These authors used it in conjunction with void ratio to describe compaction curves. In addition, Wheeler (1996) used the specific water volume $(v_w = 1+G_s w = 1+e_w)$ and the specific volume $(v = 1+e)$ as volumetric state variables describing the behavior of unsaturated soils.

3 MAIN RETENTION CURVES FOR AGGREGATED CLAYEY FABRICS UNDER CONSTANT VOLUME CONDITIONS

3.1 Tested material, equipment and procedures

Laboratory tests were conducted on artificially prepared powder (statically compacted on the dry side) obtained from a natural kaolinitic-illitic clay (20-30% kaolinite, 20-30% illite, 10-20% smectite). This moderately swelling clay has a liquid limit of $w_L = 56\%$, a plastic limit of $w_P = 29\%$, 50% of particles less than 2µm, a specific surface of $S_s = 40$ m^2/g and a specific gravity of $G_s = 2.70$. In preparing the specimens for vapor equilibrium control, powder was left in equilibrium with the laboratory atmosphere at an average relative humidity of 47% (total suction of $\psi \approx 100$ MPa) to achieve a hygroscopic water ratio of $e_w = 0.08$. For samples tested using air overpressure technique for matric suction control, the required quantity of demineralised water to achieve a predetermined $e_w = 0.40$ was added to the powder, previously cured at a relative humidity of 90%. After equalization, an initial total suction of approximately $\psi \approx 2.3$ MPa is achieved with an osmotic component of around 0.4 MPa measured by squeezing technique with a transistor psychrometer (Romero 1999). A one-dimensional static compaction procedure has been followed for both fabrication techniques until a specified final volume is achieved under a constant water ratio. Both fabrication procedures lead to an aggregated structure, as detected from freeze-dried samples compacted at different initial water ratios and observed using scanning electron microscopy. In addition, experimental results reported by Romero (1999) detected small differences in water retention characteristics of different aggregated fabrics prepared at different initial water ratios. The testing program has included different soil packings of clay aggregates, ranging from a high-porosity fabric at an initial void ratio of $e_o = 0.93$ with a tendency to collapse upon wetting to a low-porosity structure at $e_o = 0.59$ with a swelling tendency.

Experimental data of the aggregated fabrics are interpreted based on the existence of two main pore size regions, as observed from the analysis of mercury intrusion / extrusion porosimetry results (Romero et al. 1999). Firstly, an intra-aggregate porosity associated with non-constricted pores with sizes smaller than 150 nm and containing quasi-immobile water at water ratios lower than approximately 0.40. And secondly, an inter-aggregate, interconnected and constricted porosity containing free water. The term constricted porosity defines enlargements and interconnections between aggregates, which act as 'ink-bottle' necks entrapping the inter-connected porosity. Main hysteresis mechanism upon drainage and imbibition arises from these pore structure effects. Experimental results presented by Romero et al. (1999) show that intra-aggregate water represents between 54 and 59% of the total volume of water in soil in a packing with $e_o = 0.59$, whereas it corresponds to a maximum of 38% in the case of a packing compacted at $e_o = 0.93$.

Results in the high suction range were obtained from main wetting and drying paths performed with single-stage vapor equilibrium technique under free swelling or shrinkage conditions $(\sigma_m-u_a) = 0$. Compacted soil samples at different initial void ratios ranging from 0.30 to 0.90 were equilibrated at a specified relative humidity of the air. The drying paths were performed on compacted specimens that were equilibrated at a suction of 3 MPa before the following suction increase step. At the end of each single-stage equilibrium test the samples were carefully weighed and measured, and the water content of the different specimens was determined. Suction controlled oedometer and isotropic cells using air overpressure technique were employed for the study in the low suction range (under 0.5 MPa). A water volume change indicator connected to the high air-entry ceramic disc was used to obtain the values of inflow and outflow of water for each suction condition. In order to accurately determine water volume changes it was necessary to account for water volume losses due to evaporation in the open air overpressure chamber (evaporative fluxes originated due to the difference in vapor pressure between soil voids and the above air). It was also important to flush air bubbles periodically from below the high air-entry disc in order to avoid progressive cavitation of the system and the consequent loss of continuity between the pore water and the water in the measuring system, specially at high-applied matric suction values. More details concerning the experimental procedure are presented in Romero (1999).

3.2 *Test results and interpretation*

In Figure 1, the relationships between total suction and water ratio in main wetting and drying paths and at two different void ratios (isochoric conditions) have been plotted for an aggregated structure. Main wetting paths refer to the slow water absorption of an initially dry sample, while first monotonic gradual desorption starting from the minimum suction ever experienced is referred to as main drying. Data at high suction values have been interpolated at a constant void ratio from free and shrinkage swelling data. Main wetting and drying data at low suction values are obtained from constant volume swelling pressure tests using air overpressure technique. Special care has been taken to avoid that induced shrinkage upon drying could alter null volume change condition.

The hysteresis loop observed in Figure 1 and defined by the main wetting and drying boundary curves at a constant void ratio encloses the scanning region and separates attainable (inside the loop) from unattainable states for a given void ratio. Two regions can be defined in the main retention curves: an intra-aggregate water region and an inter-aggregate water region, where water ratio is high enough to partly fill the inter-aggregation voids. This last region presents a dominant capillary storage mechanism, which is dependent on void ratio and is consequently sensitive to mechanical actions. These mechanical actions produce a change in macroporosity, affecting meniscus and bulk water contained in the inter-aggregate and interconnected pores, while not greatly influencing intra-aggregate porosity. Air-entry value of the packing and pore size mechanism are the predominant factors affecting hysteresis in this inter-aggregate region.

On the other hand, at low water ratios corresponding to the intra-aggregate water (less than 0.40 as indicated in Fig. 1), the influence of the initial and equilibrated void ratios is found to be negligible, signifying that the suction - water ratio relationship is mainly dependent on the specific surface S_s of the clay and controlled by soil intra-aggregate structure and pore fluid chemistry. This microstructure, which presents an adsorption storage mechanism and contains quasi-immobile water, will be water-saturated in a main wetting path before the macropores (even so, the aggregates will continue adsorbing water after reaching their saturation). Under these conditions, a change in density of the soil basically reflects a change of the larger macroscopic voids, which have little influence on water ratio change and consequently on soil suction. Similar observations have been detected on dry-side compacted sand-bentonite mixtures by Wan et al. (1995), where experimental data clearly show that total suction can be related to the gravimetric water content even though these data come from specimens showing a wide range of initial void ratios. Delage et al. (1998a) testing compacted bentonite observed little differences with respect to gravimetric water content at high suctions between the retention curve at constant volume and the curve under free swelling conditions. The predominant hysteresis mechanism in this zone is associated with adsorbed water on clay surface mechanism, which is described in Iwata et al. (1995).

Material parameters defining the linear part in a semi-log plot of the intra-aggregate zone are indicated in Figure 1: the intersection parameter a, the slope parameter b and the microstructural water ratio e_{wm}. These parameters mainly depend on the specific surface S_s of the clay. In Figure 2, different clays, silty clays and sand-bentonite mixtures are compared, where S_s values have been estimated based on mineralogical mixture properties of the powder. Some results have been calculated from published data and others have been determined by the authors. In general, a quite good trend has been observed for the intersection and the slope parameters in terms of S_s, where empirical functions have been fitted to the data. These functions are indicated in Figure 2. However, the slope parameter is also influenced by the adsorbed water hysteresis, as well as by soil ag-

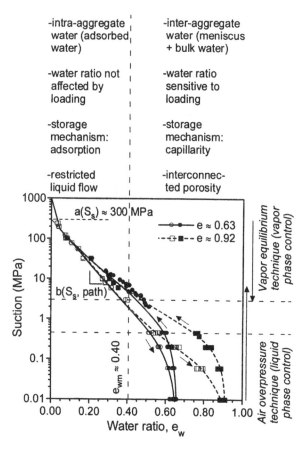

Figure 1. Main wetting and drying retention curves at constant void ratio for high porosity ($e \approx 0.92$) and low porosity ($e \approx 0.63$) specimens. Intra- and inter-aggregate zones.

gregation. Vanapalli et al. (1999) detected differences between retention curves due to the influence of initial water content affecting aggregation and soil structure. It is expected a higher resistance to desaturation in matrix dominant fabrics, which are associated with higher values of parameter b.

On the other hand, microstructural water ratios e_{wm} indicated in Figure 2 have been indirectly determined from the analysis of mercury intrusion / extrusion porosimetry results, relative water permeability values and the interpretation of contours of equal suction represented in dry density versus gravimetric water content plots, following a similar procedure to that described in Romero et al. 1999. An important aspect is that the fitted line in terms of e_{wm} and S_s, as well as the corresponding one fitting hygroscopic water ratios at a relative humidity of 47% and S_s, are parallel as observed in the plot at the right of Figure 2. This plot allows the approximate estimation of S_s for different clays based on the hygroscopic water ratio at a relative humidity of about 50%, which can be used to determine the rest of material parameters in the intra-aggregate zone. The hygroscopicity of a porous material has also been used by Keeling (1961), who proposed a methodology for S_s determination based on this concept. It is important to explain that estimations presented in Figure 2 are preliminary values and further experimental results are required to validate these data.

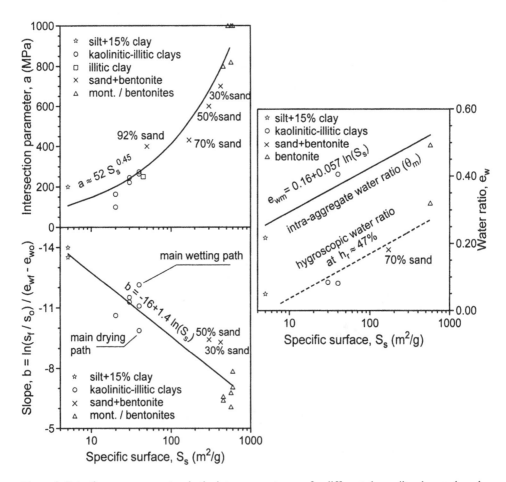

Figure 2. Retention curve parameters in the intra-aggregate zone for different clays, silty clays and sand-bentonite mixtures, as a function of specific surface.

The separation between inter- and intra-aggregate zones indicated previously is expected not to be clearly identified in very high-density packings or fabrics compacted at high water ratios with dominant matrix structure, where the inter-aggregate zone is assumed to be reduced. Under these circumstances, no important hysteresis upon suction reversal is expected (mainly induced by the adsorption hysteretic mechanism). Delage et al. (1998b) and Cui et al. (1998) observed no significant hysteresis in retention curves of heavily compacted clays, which can be explained in terms of the very dense state of the samples in which quasi-reversible intra-aggregate effects are predominant compared to inter-aggregate capillary phenomena.

Main wetting and drying paths under isochoric conditions represented in Figure 1 have been fitted to a modified form of van Genuchten (1980) expression to take into account a maximum total suction value of $a = 300$ MPa under $e_w = 0$:

$$\frac{e_w}{e} = S_r = C(s)\left[\frac{1}{1+(\alpha s)^n}\right]^m \; ; \; C(s) = 1 - \frac{\ln\left[1+\frac{s}{a_r}\right]}{\ln\left[1+\frac{a}{a_r}\right]} \quad \text{with} \quad 0.1a < a_r \le a \quad (4)$$

The original relationship with three different soil parameters $1/\alpha$ (related to the air-entry of the packing in a drying path or to the water-entry in a wetting path), n (related to the slope of the inflection point and approximately equivalent to parameter b in Fig. 1) and m (related to the residual water ratio), has been affected by the correction function $C(s)$, similar to that proposed by Fredlund & Xing (1994). The objective of this correction function is to smooth the steep behavior of the intra-aggregate region, tending to a linear relationship between the logarithm of the suction s and the water ratio e_w. Parameter a_r controls the shape and slope of the retention curve at low water ratios. Other correction factors that serve for the same purpose are given in Leong & Rahardjo (1997).

A non-linear curve-fitting algorithm using the least-squares method has been used to determine parameters n, m and α. For simplicity, $a = a_r = 300$ MPa. Fitting parameters for different aggregated fabrics ranging from $e = 0.59$ to 0.93 are indicated in Figure 3. Usually, parameter α is the one that shows more variation between the different fabrics, as well as in main wetting and drying paths for a fixed e. As observed in this figure and for simplicity, parameters n and m can be considered constant for both main wetting and drying curves at constant e.

In Figure 3, values of matric suction to reach specific degrees of saturation ($S_r = 0.90$ and 0.95) in main drying paths are also plotted for different void ratios. These values approximately define the air-entry value region of the main drying paths, associated with the maximum curvature of the retention curves indicated in Figure 1 and separating quasi-saturated from unsaturated conditions. As observed, a clear decrease of the air-entry value is detected at increasing void ratios. Ng & Pang (2000) have presented experimental data showing the tendency for a soil specimen subjected to higher stresses and associated with smaller macropore sizes, to possess a larger air-entry value. In addition, Kawai et al. (2000) have reported test results showing the variation of the air-entry value with void ratio for a silty clay. Air-entry values s_a to reach specific degrees of saturation as a function of void ratio e have been fitted to the expression indicated in Figure 3.

4 RETENTION CURVES FOR AGGREGATED CLAYEY FABRICS UNDER VARYING VOLUME CONDITIONS

4.1 Linear scaling in the inter-aggregate region at different void ratios

Approximate main water retention curves in the inter-aggregate region can be derived by linear scaling the main wetting and main drying curves for a fixed void ratio. Linear scaling of main drying water retention curves of a rigid soil has been used by van Dam et al. (1996) in order to derive a hysteretic drying scanning curve. Retention curves derived in this way may display the same air-entry value corresponding to the selected packing at a specified void ratio. Predicted retention curves for the packing at $e \approx 0.63$ and based on a reference curve at $e \approx 0.92$ are presented in Figure 4. Retention curves under isochoric conditions and in the inter-aggregate region ($e > e_{wm}$) have been fitted to a modified form of Equation 4:

$$\frac{e_w - e_{wm}}{e - e_{wm}} = S_{rM} = C(s)\left[\frac{1}{1+(\alpha s)^n}\right]^m \;;\quad C(s) = 1 - \frac{\ln\left[1+\dfrac{s}{s_m}\right]}{\ln 2} \tag{5}$$

where s_m represents the suction at e_{wm} and S_{rM} stands for the degree of saturation of the macroporosity. Material parameters for the reference packing ($e \approx 0.92$) are $e_{wm} = 0.40$, $m = 0.24$, $n = 1.17$, $s_m = 5$ MPa for main drying and $s_m = 2$ MPa for main wetting, $\alpha = 3$ MPa^{-1} for main drying and $\alpha = 30$ MPa^{-1} for main wetting, according to the results presented in Figure 3. In general, a quite good agreement is observed in Figure 4 between scaled retention curves and test results. However, greater deviations from measured values are detected in the main wetting curve, where the water-entry region of the experimental data is shifted towards higher suction values.

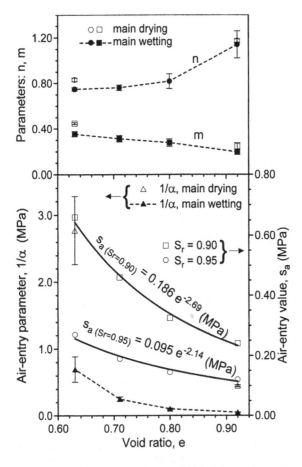

Figure 3. Material parameters (m, n and $1/\alpha$) fitting main wetting and drying retention curves for varying void ratios. Air-entry values s_a for varying void ratios in main drying paths at $S_r = 0.90$ and 0.95.

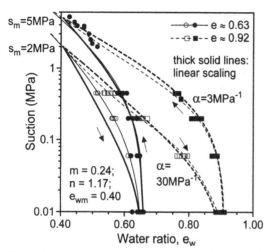

Figure 4. Scaling of retention curves in the inter-aggregate region ($e_w > e_{wm}$). Reference curve at $e \approx 0.92$. Scaled curve at $e \approx 0.63$.

4.2 Test results on low-porosity packings

Figure 5 presents isotropic test results performed on a low-porosity packing at the following initial conditions (point A): $e_o = 0.59$, $e_{wo} = 0.40$, $(u_a-u_w)_o = 1.9$ MPa and $(\sigma_m-u_a) = 0.085$ MPa. Maximum fabrication stresses, following a one-dimensional static compaction procedure at a constant water ratio, are $(\sigma_v-u_a)_{max} = 4.5$ MPa and $(\sigma_h-u_a)_{max} = 1.8$ MPa at $(u_a-u_w)_{max} = 1.9$ MPa. The lateral stress value was determined with an active lateral stress system under oedometer conditions (Romero 1999). Overconsolidation ratio, in a mean stress sense, at point A is $(\sigma_m-u_a)_{max} / (\sigma_m-u_a)_o = 32$.

The retention curve plot (upper diagram at the left in Fig. 5) of this expansive clay fabric is complemented with the stress path plot (upper diagram at the right), the volumetric state variable plane (bottom diagram at the left) and the isotropic compression plane (bottom diagram at the right). Almost all of the plots are conventional representations used in unsaturated soil mechanics, except for the volumetric state variable plane. This last diagram is equivalent to the one used in the conventional representation of compaction curves (dry density : gravimetric water content), where contour lines of equal degree of saturation can be plotted. The combined representation of experimental data in these planes allows the interpretation of retention curve results in a wider context of hydro-mechanical coupling. To the authors' knowledge, this combination of plots to represent the stress-strain evolution of an unsaturated soil is not available in the literature. This multiple diagram will be referred as SWEP plot: S for matric suction, W for water ratio, E for void ratio and P for net mean stress.

The stress paths followed (A→G) can be summarized as follows. Wetting and drying cycles (A→E), under a constant isotropic stress of $(\sigma_m-u_a) = 0.085$ MPa, were performed by maintaining a constant air pressure $u_a = 0.50$ MPa and controlling water pressures. Suction steps were applied in the wetting path up to $(u_a-u_w) = 0.01$ MPa (point B). Afterwards, the sample was subjected to a drying path up to $(u_a-u_w) = 0.45$ MPa (point C). This cycle was then repeated (C→E). Subsequently, the sample was subjected to loading (E→F) and unloading (F→G) paths at constant $(u_a-u_w) = 0.20$ MPa.

The low-porosity packing initially follows the main wetting retention curve corresponding to the isochoric curve at $e \approx 0.63$ (top diagram at the left). As the wetting process advances the path diverges towards higher values of water ratios as a result of soil swelling (bottom diagram at the right). Evolution of degree of saturation is also indicated in the figure (bottom diagram at the left), where a tendency to reach complete saturation is observed at low matric suctions. At the beginning of the wetting path, the saturation increases slightly with $\delta e / \delta e_w \approx 0.95$, but at a certain point there is a sharp increase of S_r when the macropores are flooded and the water-entry

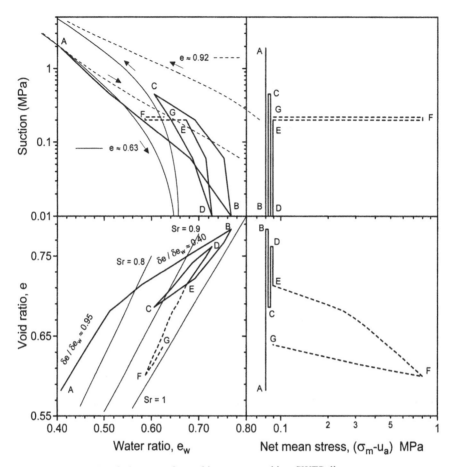

Figure 5. Test results of a low-porosity packing represented in a SWEP diagram.

value surpassed (δe / δe_w → 0.40). It appears that these important changes of S_r are associated with the occurrence of irrecoverable hydraulic processes. This way, as a first approximation, the sharp increase can be related to the dragging of the suction-decrease SD yield locus as defined by Gens (1993) upon main wetting, which is associated with an irreversible swelling of the macrostructure and the flooding of macropores. However, more experimental results are required to support this hypothesis.

Upon suction reversal in a main drying path (B→C), an initial rigid response is observed under a nearly constant degree of saturation, associated with low water and void ratio changes (δe / δe_w ≈ 0.94). However, at the air-entry zone (S_r between 0.90 and 0.95), the degree of saturation decreases sharply due to the desaturation of the largest inter-aggregate pores, expelling water as air enters the soil and affecting the macroporosity of the packing (δe / δe_w ≈ 0.42). In this suction increase path it is easier to expel water than to induce deformation on soil skeleton and S_r always reduces. It is also admitted that when matric suction increases over this air-entry zone, which bounds the transition between elastic and virgin states, both simultaneous irreversible strains and irrecoverable water ratios will develop from this suction loading affecting in a poroplastic way soil behavior. This air-entry zone, which can be associated to the suction-increase SI yield locus as defined by Alonso et al. (1990), is also expected to be modified due to the initial swelling of the macrostructure (dragging of the suction-decrease SD yield curve), originating a more open macrostructure with a lower air-entry value.

During scanning wetting (C→D), the degree of saturation increases Sr → 1, and both water and void ratios do not vary significantly. Scanning wetting refers to the first wetting path after performing the main wetting-drying cycle, even if the scanning curve does not lie between these main curves. As observed in the retention curve plot, the ending point (D) of this scanning path differs from the starting point of the main drying curve (point B) because of irreversible shrinkage aspects previously indicated, affecting the topology and water storage capacity of the macroporosity network. The simultaneous occurrence of volumetric plastic strains and plastic volumetric water content changes associated with drying paths has been mathematically formulated using a hardening poroplasticity unifying approach by Dangla et al. (1997). However, their theoretical formulation based on the physical interpretation of soil water retention curve hysteresis caused by pore structure effects was not contrasted with experimental data. Reversible water ratio changes in scanning paths are discussed in section 4.4.

The loading path (E→F) also develops simultaneous irreversible strains (bottom diagram at the right) and water ratio variations (upper diagram at the left) when activating the loading-collapse LC yield locus (Alonso et al. 1990), which are not recovered upon unloading. Aspects of irrecoverable water ratios in loading-unloading paths at constant suction are described in section 4.4.

4.3 Test results on high-porosity packings

Figure 6 presents a SWEP plot of oedometer test results performed with a lateral stress cell on a high-porosity packing at the following initial conditions (point A): $e_o = 0.88$, $e_{wo} = 0.40$, $(u_a-u_w)_o$ = 1.9 MPa, $(\sigma_v-u_a) = 0.60$ MPa and $(\sigma_h-u_a)_o = 0.20$ MPa $((\sigma_m-u_a)_o = 0.33$ MPa and $(\sigma_v-\sigma_h)_o = 0.40$ MPa). Maximum fabrication stresses, following a one-dimensional static compaction procedure at constant water ratio, are $(\sigma_v-u_a)_{max} = 1.2$ MPa and $(\sigma_h-u_a)_{max} = 0.44$ MPa at $(u_a-u_w)_{max} = 1.9$ MPa. Overconsolidation ratio, in a mean stress sense, at point A is $(\sigma_m-u_a)_{max} / (\sigma_m-u_a)_o = 2$.

Wetting (A→B and C→D) and drying (B→C and D→E) paths were performed under a constant vertical stress of $(\sigma_v-u_a) = 0.60$ MPa. Afterwards, the sample was subjected to loading (E→F and G→H) and unloading (F→G) paths at constant $(u_a-u_w) = 0.20$ MPa. Deviatoric stress $(\sigma_v-\sigma_h)$ values are indicated in the stress path plot (top diagram at the right).

The meta-stable-structured packing (with collapsible tendency upon wetting) initially follows the main wetting retention curve corresponding to the isochoric curve at $e \approx 0.92$ (top diagram at the left). As the wetting process advances, however, the path diverges towards values of water ratio lower than those corresponding to $e \approx 0.92$ packing. As observed in the volumetric state variable plane, important changes of S_r are developed in the wetting path (A→B), where the volumetric state variable path ($\delta e / \delta e_w \approx -0.62$) crosses almost perpendicularly the lines of equal S_r values. Again, these important changes of S_r are associated with the occurrence of irrecoverable processes, but in this case with a dominant mechanical feature (increase of S_r mainly due to the collapse of clay skeleton). This way, the sharp increase of S_r can be related to the dragging of the loading-collapse LC yield locus as defined by Alonso et al. (1990) upon main wetting, which is associated with an irreversible collapse of the macrostructure and the flooding of macropores. It is expected that this macrostructural strain hardening process induces a closer packing with a higher air-entry zone, which affects the subsequent activation of the suction increase SI yield locus upon main drying.

Aspects of scanning wetting and drying paths, as well as irreversible features in both volumetric state variables in loading and unloading paths, are described in section 4.4.

4.4 Scanning wetting and drying paths. Water retention characteristics in loading and unloading paths

The hysteresis loop defined by the main wetting and drying boundary curves at constant void ratio (refer to Fig. 1) encloses the scanning attainable region. This section describes scanning wetting and drying test results at different constant void ratio values, focusing on the low suction range under nearly saturated conditions ($S_r > 0.85$).

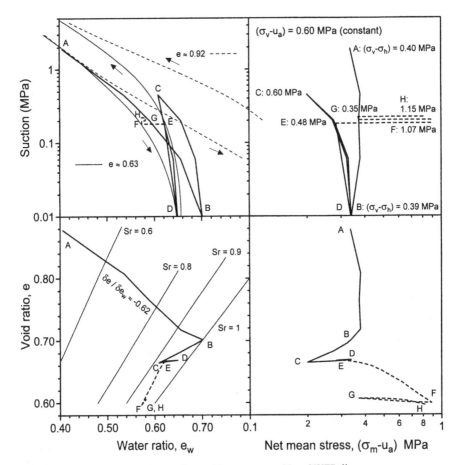

Figure 6. Test results of a high-porosity packing represented in a SWEP diagram.

As a first approximation, recoverable hydraulic behavior (indicated with superscript e) is formulated according to the following incremental relation:

$$-\delta e_w^e = \kappa_{ws}\delta s + \kappa_{wp}\frac{\delta(\sigma_m - u_a)}{(\sigma_m - u_a)} \tag{6}$$

where κ_{ws} is the scanning curve slope in the retention curve plot at constant $(\sigma_m\text{-}u_a)$ and κ_{wp} is the slope of the unloading-reloading line in a $e_w : \ln(\sigma_m\text{-}u_a)$ diagram at constant suction s (refer to Fig. 8).

Changes in recoverable water ratios associated with changes in suction under isochoric conditions are represented in Figure 7. These data have been obtained from suction controlled swelling and shrinkage pressure tests described in Romero (1999). For isochoric elastic conditions, the stress path follows:

$$\delta\varepsilon_v^e = \frac{\delta(\sigma_m - u_a)}{K_p} + \frac{\delta s}{K_s} = 0; \quad K_p = \frac{(1+e)(\sigma_m - u_a)}{\kappa}; \quad K_s = \frac{(1+e)(s + p_{at})}{\kappa_s} \tag{7}$$

where the volumetric elastic strain ε_v^e is expressed as a function of net mean stress and matric suction changes according to Alonso et al. (1990). K_p and K_s are the bulk moduli due to net stress and suction changes, where κ is the slope of the unloading-reloading line in the $e : \ln(\sigma_m-u_a)$ diagram, κ_s the slope of the reversible wetting-drying line in the $e : \ln(s)$ diagram and p_{at} the atmospheric pressure. Values of κ_s / κ are around 0.49±0.27 for the low-porosity fabric at $e \approx 0.63$ and somewhat higher 0.76±0.30 for the soil fabric at $e \approx 0.92$, according to results presented by Romero (1999). κ values are also reported in Figure 8 for both fabrics. Using Equation 7, the isochoric scanning curve slope of the retention curve $\bar{\kappa}_{ws}$ can be expressed as:

$$\bar{\kappa}_{ws} = -\left.\frac{\delta e_w}{\delta s}\right|_{e=const} = \kappa_{ws} - \frac{\kappa_s}{\kappa}\frac{\kappa_{wp}}{(s+p_{at})} \tag{8}$$

As observed in Figure 7, some dependence on e has been detected for the scanning parameter $\bar{\kappa}_{ws}$. More experimental results are required to validate this dependence. Other expressions, different from Equation 6, can be proposed as soon as more experimental data are available.

Irreversible aspects on volumetric state variables in loading and unloading paths under constant s have also been studied. Test results of both low- and high-porosity fabrics previously described (sections 4.2 and 4.3) and after a wetting and drying history, are represented in Figure 8 on a combined plot $e : (\sigma_m-u_a)$, $e_w : (\sigma_m-u_a)$ and $e_w/e : (\sigma_m-u_a)$. These volumetric state variables show clear pre- (associated with reversible processes) and post-yield zones. A common yield stress state is identified for all these state variables (yield stresses are indicated in Fig. 8 for both fabrics).

Volumetric state variable changes with net mean stress changes under constant suction are described by the following parameters: $\kappa_{wp} = -\delta e_w / \delta \ln(\sigma_m-u_a)$ (refer also to Equation 6) and $\kappa = -\delta e / \delta \ln(\sigma_m-u_a)$ in the pre-yield range, and $\lambda_w(s) = -\delta e_w / \delta \ln(\sigma_m-u_a)$ and $\lambda(s) = -\delta e / \delta \ln(\sigma_m-u_a)$ in the post-yield range, where some dependence on suction is admitted in this post-yield zone. Parameter values are also indicated in the previous figure. Values of the parameter associated with water ratio changes in the post-yield zone are slightly lower than the equivalent compressibility parameters ($\lambda_w(s) < \lambda(s)$), presenting the same value under pre-yield conditions ($\kappa_{wp} = \kappa$). This equality under pre-yield conditions is also in agreement with experimental results reported by Zakaria (1995) for isotropic unloading tests at a constant suction.

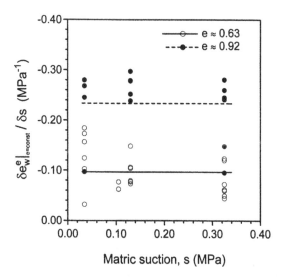

Figure 7. Changes in recoverable water ratios associated with changes in suction under isochoric conditions.

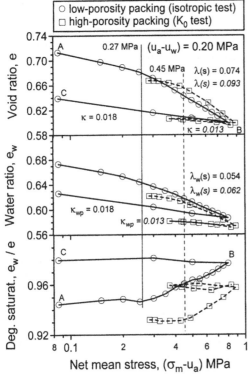

Figure 8. Loading-unloading paths at $(u_a\text{-}u_w) = 0.20$ MPa. Evolution of volumetric state variables. Values in italic indicate parameters associated with the high-porosity fabric.

The evolution of the degree of saturation $S_r = e_w/e$ exhibits an increasing trend upon loading in the post-yield zone as observed in Figure 8 and as a consequence of a higher efficiency of the loading mechanism in deforming soil skeleton than expelling water. No significant degree of saturation changes are detected in the pre-yield range.

5 SUMMARY AND CONCLUSIONS

Wetting and drying retention curves in terms of water ratio and under constant volume conditions have been presented for aggregated clay fabrics. The results indicate a delimiting zone in the retention curve separating a region of intra-aggregate porosity from an inter-aggregate porosity adjoining region. Retention curve parameters in the intra-aggregate region, which have been presented for different clays and clayey mixtures, are mainly dependent on clay specific surface. In the inter-aggregate region, water ratio is strongly coupled to mechanical effects. Deforming skeleton aspects in this inter-aggregate region have been introduced using a linear scaling via void ratio of the main retention curves for a fixed void ratio. Matric suction values to reach the air-entry zone in this inter-aggregate region have shown to depend on void ratio and on the wetting history.

Wetting and drying retention curves under varying void ratios in expansive and collapsible clay fabrics, under isotropic and oedometer conditions, have also been reported. A multiple diagram has been introduced to describe test results, which includes the following plots: retention curve plane, isotropic compression plane, stress path plane and volumetric state variable plane.

Aspects of scanning wetting and drying paths at different constant void ratios have also been reported. Irreversible aspects on volumetric state variables in loading and unloading paths under constant suction have also been studied. These volumetric state variables show clear pre- (associated with reversible processes) and post-yield zones upon loading. A common yield stress state is identified for all these volumetric state variables.

Experimental tests have presented consistent results of coupled hydro-mechanical aspects, offering an interesting challenge for future research in constitutive modeling.

REFERENCES

Alonso, E.E., A. Gens & A. Josa 1990. A constitutive model for partially saturated soils. *Géotechnique* 40(3): 405-430.

Aubertin, M., J.-F. Ricard & R.P. Chapuis 1998. A predictive model for the water retention curve: application to tailings from hard-rock mines. *Can. Geotech. J.* 35:55-69.

Barbour, S.L. 1998. Nineteenth Canadian Geotechnical Colloquium: The soil-water characteristic curve: a historical perspective. *Can. Geotech. J.* 35: 873-894.

Biarez, J., J.M. Fleureau, M.-I. Zerhouni & B.S. Soepandji 1988. Variations de volume des sols argileux lors de cycles de drainage-humidification. *Rev. Franç. Géotech.* 41: 63-71.

Chen, Z.-H., D.G. Fredlund & J. K.-M. Gan 1999. Overall volume change, water volume change, and yield associated with an unsaturated compacted loess. *Can. Geotech. J.* 36: 321-329.

Cui, Y.G., M. Yahia-Aissa & P. Delage 1998. A model for the volume change behaviour of heavily compacted swelling clays. *Proc. 5th Int. Workshop on Key Issues in Waste Isolation Research, Barcelona, 2-4 December 1998.*

Dangla, P., L. Malinsky & O. Coussy 1997. Plasticity and imbibition-drainage curves for unsaturated soils: A unified approach. In Pietruszczak & Pande (eds), *Proc. 6th Int. Symp. on Numerical Models in Geomechanics, Montreal, 2-4 July 1997*: 141-146. Rotterdam: Balkema.

Delage, P., Y.J. Cui, M. Yahia-Aissa & E. De Laure 1998a. On the saturated hydraulic conductivity of a dense compacted bentonite. *Proc. 2nd Int. Conf. on Unsaturated Soils, Beijing, 27-30 August 1998*, 1: 344-349. Beijing: International Academic Publishers.

Delage, P., M.D. Howat & Y.J. Cui 1998b. The relationship between suction and swelling properties in a heavily compacted unsaturated clay. *Engineering Geology* 50: 31-48.

Dineen, K. & A.M. Ridley 1999. The soil moisture characteristic curve. *Proc. 11th Panamerican Conf. on Soil Mechanics and Geotechnical Engineering, Foz do Iguaçu, 8-12 August 1999*: 1013-1018.

Edgar, T.V. 1993. One and three dimensional, three phase deformation in soil. In S.L. Houston and W.K. Wray (eds). *Unsaturated Soils Geotechnical Special Publications N° 39, Dallas*: 139-150. Philadelphia: ASTM.

Fleureau, J.-M., S. Kheirbek-Saoud, R. Soemitro & S. Taibi 1993. Behaviour of clayey soils on drying-wetting paths. *Can. Geotech. J.* 30: 287-296.

Fredlund, D.G. & A. Xing 1994. Equations for the soil-water characteristic curve. *Can. Geotech. J.* 31: 521-532.

Gens, A. 1993. Constitutive modelling of expansive soils. *Unsaturated soils: Recent developments and applications, Civil Engineering European Courses, Barcelona*, 15-17 June 1993.

Houlsby, G.T. 1997. The work input to an unsaturated granular material. *Géotechnique* 47(1): 193-196.

Iwata, S., T. Tabuchi & B.P. Warkentin 1995. *Soil-water interactions. Mechanisms and applications.* New York: Marcel Dekker, Inc., 2nd ed.

Kaczmarek, M. & T. Hueckel 1998. Use of porosity in models of consolidation. *J. Engrg. Mechanics ASCE* 124(2): 237-239.

Kawai, K., S. Kato & D. Karube 2000. The model of water retention curve considering effects of void ratio. In H. Rahardjo, D.G. Toll and E.C. Leong (eds.). *Proc. of the Asian Conf. on Unsaturated Soils, Singapore, 18-19 May2000*: 329-334. Rotterdam: Balkema.

Keeling, P.S. 1961. The examination of clays by IL/MA. *Transactions of The British Ceramic Society* 60: 217-244.

Leong, E.C. & H. Rahardjo 1997. Review of soil-water characteristic curve equations. *J. Geotech. Engrg. ASCE* 123(12): 1106-1117.

Ng, C.W.W & Y.W. Pang 2000. Influence of stress state on soil-water characteristics and slope stability. *J. Geotech. Engrg. ASCE* 126(2): 157-166.

Prashanth, J.P., P.V. Sivapullaiah & A. Sridharan 1998. Compaction curves on volume basis. *Geotechnical Testing Journal GTJODJ* 21(1): 58-65.

Rampino, C., C. Mancuso & F. Vinale 1999. Laboratory testing on an unsaturated soil: equipment, procedures and first experimental results. *Can. Geotech. J.* 36: 1-12.

Romero, E. 1999. *Characterization and thermo-hydro-mechanical behavior of unsaturated Boom clay: an experimental study.* PhD Thesis, Universidad Politécnica de Cataluña.

Romero, E., A. Gens & A. Lloret 1999. Water permeability, water retention and microstructure of unsaturated Boom clay. *Engineering Geology* 54: 117-127.

Toll, D. G. 1995. A conceptual model for the drying and wetting of soil. In E.E. Alonso and P. Delage (eds.). *Proc. 1st Int. Conf. on Unsaturated Soils, Paris, 6-8 September 1995*, 2: 805-810. Rotterdamm: Balkema / Presses des Ponts et Chaussées.

Vanapalli, S.K., D.G. Fredlund & D.E. Pufahl 1999. The influence of soil structure and stress history on the soil-water characteristics of a compacted till. *Géotechnique* 49(2): 143-159.

van Dam, J.C., J.H.M. Wösten & A. Nemes 1996. Unsaturated soil water movement in hysteretic and water repellent field soils. *Journal of Hydrology* 184: 153-173.

van Genuchten, M.Th. 1980. A closed-form equation for predicting the hydraulic conductivity of unsaturated soils. *Soil Sci. Soc. Am. J.* 44: 892-898.

Vaunat, J., E. Romero & C. Jommi 2000. An elastoplastic hydro-mechanical model for unsaturated soils. *Proc. International Workshop on Unsaturated Soils: Experimental Evidence and Theoretical Approaches, Trento, 10-12 April 2000*. This issue. Rotterdam: Balkema.

Wan, A.W.L., M.N. Gray & J. Graham 1995. On the relations of suction, moisture content and soil structure in compacted clays. In E.E. Alonso and P. Delage (eds.). *Proc. 1st Int. Conf. on Unsaturated Soils, Paris, 6-8 September 1995*, 1: 215-222. Rotterdamm: Balkema / Presses des Ponts et Chaussées.

Wheeler, S.J. & V. Sivakumar 1995. An elasto-plastic critical state framework for unsaturated soil. *Géotechnique* 45(1): 35-53.

Wheeler, S.J. 1996. Inclusion of specific water volume within an elasto-plastic model for unsaturated soil. *Can. Geotech. J.* 33: 42-57.

Wheeler, S.J. & D. Karube 1996. Constitutive modelling. In E.E. Alonso and P. Delage (eds.). *Proc. 1st Int. Conf. on Unsaturated Soils, Paris, 6-8 September 1995*, 3: 1323-1356. Rotterdam: Balkema / Presses des Ponts et Chaussées.

Zakaria, I. 1995. *Yielding of unsaturated soil*. PhD Thesis, University of Sheffield.

Theoretical approaches

Experimental Evidence and Theoretical Approaches in Unsaturated Soils, Tarantino & Mancuso (eds)
© *2000 Taylor & Francis, ISBN 90 5809 186 4*

Inclusion of hydraulic hysteresis in a new elasto-plastic framework for unsaturated soils

M.S.R.Buisson & S.J.Wheeler
Department of Civil Engineering, University of Glasgow, UK

ABSTRACT: Experimental results from cycles of wetting and drying on unsaturated compacted clays show forms of irreversible volume change that cannot be predicted by conventional elasto-plastic models for unsaturated clays. These additional forms of irreversible behaviour are attributed to the occurrence of hydraulic hysteresis in the variation of degree of saturation. A new form of elasto-plastic model, incorporating the influence of hydraulic hysteresis, is presented in qualitative form. The proposed modelling framework is able to predict either net swelling or net shrinkage over a cycle of wetting and drying (depending on the previous history of compaction and variation of net stress and suction).

1 INTRODUCTION

Alonso et al. (1987) presented an elasto-plastic framework to describe the behaviour of unsaturated soils, including the possibility of either swelling or collapse compression on wetting. This qualitative framework, which employed matric suction and net stress (the difference between total stress and air pressure) as stress state variables, was subsequently developed to a full elasto-plastic constitutive model by Alonso et al. (1990). Subsequent refinements and variations of the model were suggested by authors such as Wheeler & Sivakumar (1995). These models represent both swelling on wetting and shrinkage on drying as elastic processes, whereas collapse compression on wetting is modelled as a plastic process (corresponding to expansion of a Loading Collapse (*LC*) yield curve). Hence, during cycles of wetting and drying over a constant suction range at a constant value of net stress, the models predict that the only possible source of irreversible volume changes is if plastic collapse compression occurs during the first wetting path. Experimental evidence, from a variety of sources, indicates, however, additional forms of irreversible volume change during cycles of wetting and drying, which cannot be represented by elasto-plastic models of the type first proposed by Alonso et al. (1990).

2 IRREVERSIBLE VOLUME CHANGES DURING CYCLES OF WETTING AND DRYING.

Chu & Mou (1973) reported experimental test results from cycles of wetting and drying performed on a highly expansive clay, where a large irreversible component of swelling was observed during the first wetting path (fig.1). Subsequently, Alonso et al. (1995) presented controlled-suction oedometer test results for another highly expansive clay, showing an irreversible component of shrinkage during the drying stages of wetting and drying cycles (fig. 2). During the first wetting path C1, from a very high initial value of suction, initial swelling was followed by some collapse compression as the suction was progressively reduced. The subsequent cycles of wetting and drying showed a significant component of irreversible shrinkage during drying paths C2 and C4, with a clear suggestion of a yield point in each drying path. The magnitude of the irreversibility was greater in the first drying path C2 than in the second drying path C4. The types of behaviour reported by Chu & Mou (1973) and Alonso et al. (1995), illustrated in figures 1 and 2, cannot be represented by a constitutive model of the type presented by Alonso et al. (1990).

In the light of experimental observations of irreversible volume changes in addition to those attributable to collapse compression on wetting, Gens & Alonso (1992) and later Alonso et al (1994) presented a modified model for unsaturated highly expansive clays. In this new modelling framework, irreversible swelling on wetting (such as reported by Chu & Mou (1973)) or irreversible shrinkage during drying (such as reported by Alonso et al. (1995)) were attributed to elastic swelling or compression of the saturated microstructure of individual clay packets exceeding a maximum value that could be tolerated without plastic re-arrangement of the unsaturated macrostructure. In the proposed elasto-plastic framework, this was represented by the inclusion of Suction Decrease (*SD*) and Suction Increase (*SI*) yield curves.

Sharma & Wheeler (2000) questioned the validity of the model for highly expansive clays proposed by Gens & Alonso (1992) and Alonso et al. (1994), in the light of the experimental evidence from Chu & Mou (1973), Alonso et al. (1995) and more recent laboratory test data from Sharma (1998). From experimental evidence they observed the following:

-Net shrinkage over a wetting-drying cycle and net swelling over a wetting-drying cycle were observed by Sharma (1998) on samples of the same soil, and whether net swelling or net shrinkage occurs is a function of the history of compaction pressure and the subsequent variation of net stress and suction, rather than being determined solely by soil type.

-Net swelling or net shrinkage over a cycle of wetting and drying was observed by Sharma (1998) on unsaturated pure speswhite kaolin (a non expansive soil), suggesting that irreversible swelling or shrinkage is a feature of behaviour of all unsaturated compacted clays, rather than being restricted to highly expansive clays.

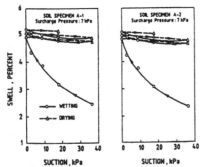

Figure 1. Wetting and drying cycles on a highly swelling soil (after Chu & Mou (1973)).

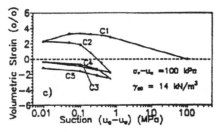

Figure 2. Wetting and drying cycles performed on Boom clay (after Alonso et al. (1995)).

-Net shrinkage over a cycle of wetting and drying occurred with evidence of yielding on the drying path but not on the wetting path. Also net swelling over a cycle of wetting and drying occurred without any evidence of yielding during either the drying or wetting paths. For the Gens & Alonso (1992) and Alonso et al. (1994) class of model, prediction of net swelling over a cycle of wetting and drying or prediction of net shrinkage over more than the first cycle would require the presence of yield points on both wetting and drying paths.

Sharma & Wheeler (2000) proposed an alternative explanation for the observed behaviour, linking irreversible volume changes during cycles of wetting and drying to the observation of hydraulic hysteresis in the variation of degree of saturation. This explanation is based on consideration of the differences in behaviour of water-filled voids and air-filled voids, and the response of a void on flooding or emptying of water during a wetting or drying path. The volume change behaviour of water-filled voids is likely to be largely controlled by the stress variable $p - u_w$. In contrast, the stress variable $p - u_a$ is relevant for the behaviour of air-filled voids, but in addition the volume changes of these air-filled voids are also affected by the presence of lenses of "meniscus water" at the surrounding inter-particle or inter-packet contacts (Wheeler & Karube (1996)). The analysis of the contact between two idealised spherical particles (see Fisher (1926)) indicates that the value of additional inter particle normal force caused by a lens of meniscus water changes as suction increases. To a first approximation, however, the additional inter-particle normal force due to meniscus water can be assumed to be constant, provided the surrounding voids remain air-filled and the meniscus water lens is present, but the additional force disappears if the voids on either side of the particle contact flood with water. Loss of this component of force can have two effects: it will cause a component of swelling due to the elastic recovery of the particles under the reduced force, but it will also make the frictional contact more susceptible to slippage (introducing the possibility of wetting-induced collapse compression).

According to Sharma & Wheeler (2000), two elements of swelling can be expected on wetting at constant $p - u_a$. One is an elastic increase in volume of water-filled voids due to the decrease in the value of $p - u_w$ and the other component is an increase in the volume of those voids that flooded with water during the wetting process and lost the additional inter-particle force arising from meniscus water. The second component arises from elastic deformation of particles or packets but is not necessarily reversible if the suction is increased again. It depends on the relevant voids emptying of water once more, rather than being dependent on the increase in suction. If there is hydraulic hysteresis in the water-retention relationship, so that the reduction of S_r during a drying path is less than the corresponding increase of S_r during the preceding wetting path, then the magnitude of shrinkage during drying will be less than the previous swelling during wetting because the second component of volume change will be absent or reduced. This would lead to net swelling over a cycle of wetting and drying, with no evidence of yielding during either wetting or drying. This is entirely consistent with the type of behaviour reported by, for example, Chu & Mou (1973)(see fig. 1).

Hydraulic hysteresis in the variation of degree of saturation also explains the possibility of yielding during a drying path, which could lead to net shrinkage over a cycle of wetting and drying. Hydraulic hysteresis means that some voids that were air-filled at a given value of suction during a wetting path will be water-filled at the same value of suction during the subsequent drying path. These voids will now be experiencing a value of $p - u_w$ that they have never previously been subjected to while water-filled and this may be sufficient to cause yielding. During the previous wetting, when the suction was of the same magnitude, these voids were experiencing a lower stress $p - u_a$ and had the stabilising effect of meniscus water at particle contacts. This mechanism for yielding during a drying path and hence net shrinkage over a cycle of wetting and drying, described by Sharma & Wheeler (2000), is entirely consistent with the type of behaviour reported by, for example, Alonso et al. (1995) (see fig. 2).

3 ELASTO- PLASTIC MODELLING OF HYDRAULIC HYSTERESIS.

If the presence of hydraulic hysteresis explains the occurrence of irreversible volume changes during cycles of wetting and drying, then it is important to understand and model the process of hydraulic hysteresis.

The physical explanation for hydraulic hysteresis developed here is based on consideration of the mechanisms governing the flooding and emptying of voids. In the interest of simplicity, in developing this explanation the soil skeleton is at first considered to be rigid. Figure 3 shows a void, during a wetting path when about to be flooded with water (a), and on a drying path when about to empty of water (b).

On wetting (fig. 3a) the radius of the menisci separating air and water increases as the suction decreases, and the menisci migrate inwards from the narrow throats (1) at the entries to an air-filled void to a new position (2). As the suction decreases further and the radius of the menisci continues to increase, a limiting condition is reached (3) when the radius is equal to that of the largest sphere that can fit within the void. Any further reduction of suction would result in the void flooding with water (Fredlund (1976)). The value of suction at which this void floods with water during a wetting path is therefore governed by the critical radius r_3 shown in figure 3a.

On drying (fig. 3b), the radius of the menisci separating air and water decreases and the menisci migrate inwards from an initial position (4) towards the narrowest point of the throats. A critical point is reached when the radius is just sufficient to bridge the largest throat joining the water-filled void to an adjacent air-filled void (5). Any further increase of suction would result in the void filling with air. The value of suction at which the void empties of water during a drying path is therefore governed by the critical radius r_5 shown in figure 3b.

As the critical radius for emptying of a void r_5, is smaller than the critical radius for flooding a void r_3, the water retention curve will inevitably exhibit hydraulic hysteresis (fig. 4), with the suction corresponding to a given degree of saturation being higher on a drying path s_b than on a wetting path s_a.

The form of the water retention relationship described above can be represented by an elasto-plastic model, with changes of degree of saturation divided into elastic (reversible) and plastic (irreversible) components.

Elastic changes of S_r are attributable to small reversible movements of the menisci separating air and water, such as between positions 1,2 and 3 in figure 3a or between positions 4 and 5 in figure 3b (in fact these menisci movements may not be entirely reversible, because of differences in the contact angle during wetting and drying). In contrast, the process of flooding or emptying a void with water produces an irreversible (plastic) component of the change of S_r. This representation of the water retention behaviour by an elasto-plastic relationship is very convenient if it is to subsequently form part of a full elasto-plastic hydro-mechanical model for a deformable unsaturated soil (see next section).

Figure 5 shows a simplified representation of hydraulic hysteresis in a rigid soil, based on a classical elasto-plastic approach. The primary drying and wetting curves shown in the figure represent drying from a fully saturated state or wetting from a dry state. Any cycle of wetting and drying has to be within these two limiting curves. Considering a complete cycle of drying and wetting 1,2,3,4,1 the variation of degree of saturation during the drying path 1,2,3 initially

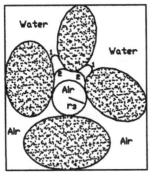 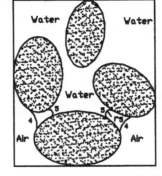

a) Void about to be filled with water.   b) Void about to be emptied of water.

Figure 3. Mechanism for flooding and emptying of a void.

shows only elastic changes until the suction reaches the air entry value for those water-filled voids that have the largest entry throats from adjacent air-filled voids, at which point plastic changes start to occur (2). On reversal of suction, from 3 to 1, again the degree of saturation experiences elastic changes until the yield value is reached at (4).

Yield on drying (2 in fig. 5) corresponds to a Suction Increase (SI) yield curve in the $s : p - u_a$ plane (see fig. 6) whereas yield on wetting (4 in fig. 5) corresponds to a Suction Decrease (SD) yield curve in figure 6. Figure 6a represents the drying path 1,2,3 and figure 6b the wetting path 3,4,1. During section 2,3 of the drying path, as the SI curve is pushed upwards there must be coupled upward movement of the SD curve, so that yield can occur at (4) during the subsequent wetting path. Similarly downward movement of the SD curve during section 4,1 of the wetting path must cause coupled downward movement of the SI curve. The form of coupling between SI and SD yield curves is described by the shape of the primary drying and wetting curves in figure 5.

Simulating the hydraulic hysteresis with a classical type of elasto-plastic formulation, as shown in figure 5, inevitably means that the modelling of hydraulic hysteresis is relatively crude. For example, continuous cycles of drying and wetting over a fixed range of suction can exhibit only either an elastic response or a closed loop spanning between the primary wetting and drying curves (whereas in reality more complex behaviour is observed). A more realistic modelling could be produced by, for example, the use of a bounding surface elasto-plastic formulation (Dafalias & Herman (1982)), but in the interests of simplicity the current discussion is limited to the classical form of elasto-plasticity illustrated in figure 5.

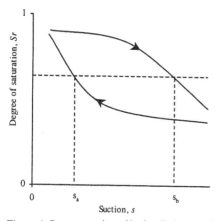

Figure 4. Representation of hydraulic hysteresis.

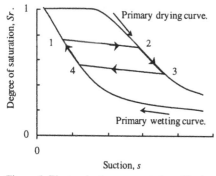

Figure 5. Elasto-plastic representation of hydraulic hysteresis in a rigid soil.

113

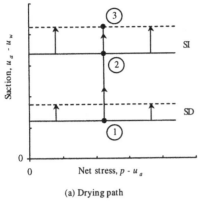
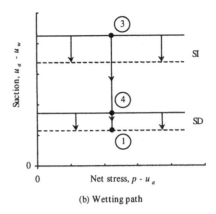

| (a) Drying path | (b) Wetting path |

Figure 6. *SI* and *SD* yield curves for a rigid soil.

4 COMBINED HYDRO-MECHANICAL ELASTO-PLASTIC MODEL.

Having modelled hydraulic hysteresis, the next step is to combine it with the stress-strain behaviour of a deformable unsaturated soil in a cross-coupled model. If volume changes of air-
filled voids and water-filled voids are dependent on different stress variables, then the mechanical behaviour will inevitably be influenced by the hydraulic hysteresis. Conversely, volumetric strains will be accompanied by changes in the size of voids and the dimensions of the
throats forming the entries to voids, and therefore deformation will inevitably influence the
water retention curve.

For the axisymmetric conditions of the triaxial test, the stress variables used in the proposed
hydro-mechanical elasto-plastic framework are mean net stress $p - u_a$, deviator stress q and
suction $u_a - u_w$, i.e. the same as in the conventional model of Alonso et al. (1990). The strain increment variables which are work conjugate to these stresses, are derived by Houlsby (1997) as
respectively the volumetric strain increment $d\varepsilon_v$, shear strain increment $d\varepsilon_s$ and a strain increment variable defined by :

$$d\varepsilon_{vw} = -\frac{dv_w}{v} \tag{1}$$

where the specific water volume v_w is defined by :

$$v_w = 1 + S_r e \tag{2}$$

This paper is limited to qualitative description of the proposed modelling framework for isotropic stress states only. The relevant stress variables are therefore mean net stress $p - u_a$ and
suction $u_a - u_w$, and the corresponding strain increment variables are $d\varepsilon_v$ and $d\varepsilon_{vw}$.

Volumetric strains are divided into elastic and plastic components, with plastic volumetric
strains associated with slippage at inter-particle or inter-packets contacts. Volume changes of
air-filled voids and water-filled voids are considered separately. Onset of plastic strains of the
air-filled voids corresponds to a Compression of Air voids (*CA*) yield curve in the $s : p - u_a$
plane (fig. 7). The *CA* yield curve is at constant $p - u_a$, because the behaviour of air-filled voids
is controlled by the stress variable $p - u_a$ and the presence of additional inter-particle normal
forces due to meniscus water that are constant in magnitude independent of the value of suction. Yielding of air-filled voids leads to plastic volumetric strains $d\varepsilon_v^p$, but no plastic changes
in specific water volume ($d\varepsilon_{vw}^p = 0$) and hence the flow rule on the *CA* yield curve is associated
(see fig. 7). Yielding on the *CA* curve will generally correspond to compression of the largest of
the air-filled voids as these are likely to be the least stable.

114

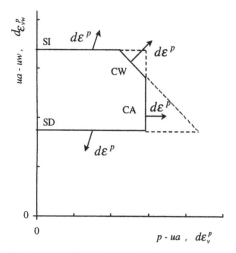

Figure 7. Yield curves for isotropic stress states.

Onset of plastic volume changes of water-filled voids corresponds to a Compression of Water voids (CW) yield curve (see fig. 7). The CW yield curve is at a constant $p - u_w$,because the behaviour of water-filled voids is controlled by the stress variable $p - u_w$. The CW curve is therefore inclined at 45° in the $s : p - u_a$ plane. Yielding of water filled voids gives rise to a plastic change of v_w that is equal in magnitude to the plastic change of specific volume. For yielding on the CW curve, the relationship between plastic strain increments is therefore $d\varepsilon_{vw}^p = d\varepsilon_v^p$, which again corresponds to an associated flow rule. Yielding on the CW curve will generally correspond to compression of the largest of the water-filled voids, because these are likely to be the least stable, but these are still likely to be smaller than all of the air-filled voids.

Depending upon the previous history of net stress and suction variation, the CW yield curve can lie entirely outside the CA yield curve (fig. 8a), entirely inside the CA yield curve (fig. 8b) or the two yield curves can intersect (fig. 7).

The flow rules for yielding on the SI or SD curves are non-associated (see fig. 7), because there are plastic (irreversible) components of volume change associated with emptying or flooding of voids with water, due to the gain or loss of the additional component of inter-particle force due to meniscus water. As described earlier, these volume changes arise from elastic deformation of individual particles or packets, but the overall process is plastic, because the volume changes are not immediately recovered on reversal of suction.

Coupling between some of the different yield curves is an important feature of the proposed framework. Movements of the SI and SD curves are cross-coupled, as described in the previous section. Movement of the SI or SD curve, corresponding to a void emptying or flooding with water, also introduces a strong coupled movement of the CW curve. If the SI curve is pushed

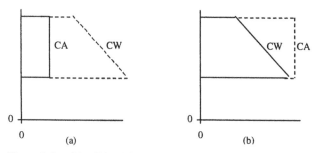

Figure 8. Possible CW and CA yield curves positions.

115

upwards by an increase in suction, this correspond to some water-filled voids becoming air-filled voids. It will typically be the largest of the water-filled voids that drains. As these are likely to be the least stable and most susceptible to yielding of the water-filled voids, the loss of them is represented by an outward movement of the *CW* curve. There is no corresponding inwards movement of the *CA* curve, because the new air-filled voids are likely to be the smaller, and hence least susceptible to yielding, than the pre-existing air-filled voids. Similarly, if the *SD* curve is pushed downwards by a decrease of suction, representing the flooding of some of the smallest air-filled voids to form some of the largest water-filled voids, there will be a coupled inward movement of the *CW* curve and no movement of the *CA* curve.

Yielding on the *CA* or *CW* curve is also likely to cause some coupled upward movement of the *SD* and *SI* curves, as plastic compression of the voids occurs, with a corresponding decrease in the dimensions of both the voids and the entry throats, and hence an increase in the suctions required to cause flooding with water during a wetting path or emptying of water during a drying path.

5 MAIN FEATURES OF BEHAVIOUR DURING WETTING AND DRYING.

The new modelling framework described above is capable of representing, at a qualitative level, not only the classical patterns of swelling and collapse compression on wetting but also the additional forms of irreversible behaviour during cycles of wetting and drying reported in section 2.

Predicted behaviour during wetting at a low value of $p - u_a$ is shown in figure 9. The stress path and yield curve positions are shown in figure 9a, with the initial locations of the *SI*, *SD* and *CW* yield curves shown by solid lines and the final locations shown by dashed lines (the *CA* yield curve has been omitted for clarity). The predicted volume changes are shown in figure 9b. Initially, during section 1-2 of the wetting path while the stress path remains inside the yield curves, only elastic swelling occurs, but when the *SD* yield curve is reached at 2, additional plastic components of swelling occur during section 2-3 of the wetting path. This additional plastic component of swelling is attributable to an increase of those voids that flooded with water during the wetting process and lost the additional inter-particle force arising from meniscus water (as described earlier). The physical explanation for the occurrence of a plastic component of swelling on wetting is therefore different to that provided by Gens and Alonso (1992). If the stress state was already on the *SD* yield curve at the start of the wetting path, due for example to previous wetting from a higher suction, no yield point would be apparent during wetting. As the *SD* yield curve is pushed down, coupled downward movement of the *SI* curve and inward movement of the *CW* curve occur, but in the example shown in figure 9 the stress path remains well inside the *CW* curve.

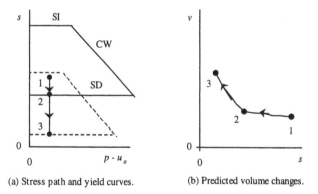

(a) Stress path and yield curves.   (b) Predicted volume changes.

Figure 9. Swelling on wetting at low value of $p - u_a$

116

Figure 10 shows the predicted behaviour during wetting at a high value of $p - u_a$. In figure 10a the initial stress state, at point 1, is assumed to be already on the SD curve, and again the CA curve has been omitted for clarity. During wetting from 1 to 2 elastic and plastic components of swelling occur, and the downward movement of the SD curve causes coupled downward movement of the SI curve and inward movement of the CW curve (to new locations shown by the chain dotted line). At point 2, the inward movement of the CW curve has been sufficient to bring the corner between SD and CW curves over the stress point. As wetting proceeds from 2 to 3 there is continued coupled inward movement of the CW curve, but this must be partially offset by directly pushing out the CW curve, to ensure that the stress point does not fall outside the yield curve (final positions of the yield curves are shown by the dashed line). This process of pushing out the CW curve corresponds to plastic collapse compression (see fig. 10b) and is associated with volume reduction of the newly water-filled void. It is useful to note that during section 1-2 of the wetting path, the trace followed by the corner between the SD and CW curves, as the CW curve is dragged inwards, corresponds to the shape of the LC yield curve in the conventional model of Alonso et al. (1990).

Figure 11 shows how the new framework is able to predict net swelling over a cycle of wetting and drying. During wetting path 1-2 both elastic and plastic components of swelling occur, whereas during drying path 2-3 only the elastic component is recovered (provided the SI curve is not reached during drying). Subsequent cycles of wetting and drying would simply re-trace the elastic path 2-3. This is the form of behaviour reported by Chu & Mou (1973)(see fig. 1).

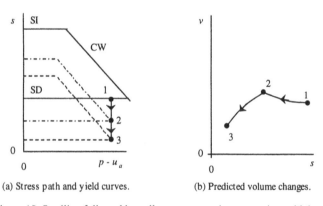

(a) Stress path and yield curves.   (b) Predicted volume changes.

Figure 10. Swelling followed by collapse compression on wetting at high value of $p - u_a$.

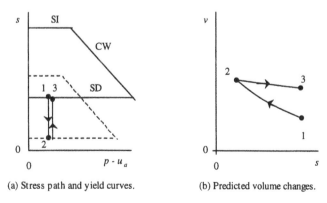

(a) Stress path and yield curves.   (b) Predicted volume changes.

Figure 11. Net swelling over a wetting and drying cycle.

117

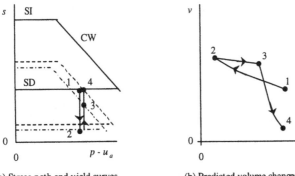

(a) Stress path and yield curves. (b) Predicted volume changes.

Figure 12. Net shrinkage over a cycle of wetting and drying.

Figure 12 shows how the new framework is able to predict net shrinkage over a cycle of wetting and drying. During wetting path 1-2 elastic and plastic components of swelling occur. As the *SD* yield curve is pushed downwards during wetting, a coupled inward movement of the *CW* curve occurs to a new position shown by the chain-dotted line. This means that during the subsequent drying path 2-3-4, it is possible for the stress path to hit the *CW* curve at point 3. From 3 to 4 the *CW* curve is pushed out, to a final position shown by the dashed line, producing a plastic component of compression during drying. This is the form of behaviour reported by Alonso et al. (1995)(see fig. 2), and also reported by Sharma & Wheeler (2000).

6 CONCLUSION

A new form of elasto-plastic model for unsaturated soil, incorporating the influence of hydraulic hysteresis, has been presented in qualitative form for the simplified case of isotropic stress states. The proposed model framework includes 4 yield curves. Suction Increase (*SI*) and Suction Decrease (*SD*) yield curves are associated with emptying or flooding of voids with water during drying or wetting respectively. A Compression of Water voids (*CW*) yield curve is associated with plastic compression of water-filled voids, arising from slippage of the surrounding inter-particle or inter-packets contacts. Similarly, a Compression of Air voids (*CA*) yield curve is associated with plastic compression of air-filled voids.

The proposed framework is able to represent, at a qualitative level: swelling or wetting at low values of $p - u_a$; collapse compression on wetting at higher values of $p - u_a$; net swelling over a cycle of wetting and drying ; and net shrinkage over a cycle of wetting and drying. In addition, Buisson (in press) shows that it is capable of predicting correctly the influence of a wetting-drying cycle on the subsequent compression behaviour during isotropic loading (as reported by Sharma (1998) and Buisson (in press)).

Work currently in progress is attempting to develop the new framework to a full elasto-plastic mathematical model.

The current version of the framework is based on classical elasto-plasticity. This results in a somewhat crude representation of the form of hydraulic hysteresis in the water retention relationship. Bounding surface plasticity could be used to provide a more realistic representation of hydraulic hysteresis.

REFERENCES

Alonso E.E., A. Gens & D.W. Hight 1987. Special problems soils. General report. *Proc. 9th Eur. Conf. Soil Mech, Dublin* 3: 1087-1146.
Alonso, E.E., A. Gens & A. Josa 1990. A constitutive model for partially saturated soils. *Géotechnique* 40(3): 405-430.

Alonso, E.E., A. Gens & W.Y.Y. Gehling 1994. Elasto-plastic model for unsaturated expansive soils. *Proc. 3rd Eur. Conf. Num. Methods Geotech. Eng., Manchester*: 11-18.

Alonso, E.E., A. Lloret, A. Gens & D.Q. Yang 1995. Experimental behaviour of highly expansive double-structure clay. *Proc. 1st Int. Conf. Unsaturated Soils, Paris* 1: 11-16.

Buisson, M.S.R. (in press). Mechanical behaviour of unsaturated compacted clays during stress paths involving wetting and drying cycles. *PhD thesis*, University of Glasgow, U.K.

Chu, T.Y. & C.H. Mou 1973. Volume change characteristics of expansive soils determined by controlled suction tests. *Proc. 3rd Int. Conf. Expansive Soils, Haifa* 1: 177-185.

Dafalias, Y.F. & L.R. Herman 1982. Bounding surface formulation of soil plasticity. In G. N. Pande. and O. C Zienkiewicz (eds.), *Soil Mechanics - Transient and cyclic loads:* 252-282.

Fisher, R.A. 1926. On the capillary forces in an ideal soil; correction of formulae given by W.B Haines. *J. Agri. Sci.* 16: 492-505.

Gens, A. & E.E. Alonso 1992. A framework for the behaviour of unsaturated expansive clays. *Can. Geotech. J.* 29: 1013-1032.

Fredlund, D.G. 1976. Density and compressibility characteristics of air-water mixtures. *Canadian Geotechnical Journal* 13 (4): 386-396.

Houlsby, G.T. 1997. The work input to an unsaturated granular material. *Géotechnique* 47(1): 193-196.

Sharma, R.S. 1998.Mechanical behaviour of unsaturated highly expansive clays. *DPhil thesis*, University of Oxford, U.K.

Sharma, R.S & S.J. Wheeler 2000. Behaviour of an Unsaturated Highly Expansive Clay during Cycles of Wetting and Drying. *Proc. Asian Conf. On Unsaturated soils, Singapore*: 721-726.

Wheeler, S. J. & D. Karube 1996. State of the art report - Constitutive modelling. *1st Int. Conf. on Unsaturated soils, Paris* 3: 1323-1356.

Wheeler, S. J. & V. Sivakumar 1995. An elasto-plastic critical state framework for unsaturated soil. *Géotechnique* 45(1): 35-53.

Experimental Evidence and Theoretical Approaches in Unsaturated Soils, Tarantino & Mancuso (eds)
© 2000 Taylor & Francis, ISBN 90 5809 186 4

An elastoplastic hydro-mechanical model for unsaturated soils

J. Vaunat & E. Romero
Departamento de Ingeniería del Terreno, Cartográfica y Geofísica, Universidad Politécnica de Cataluña, Barcelona, Spain

C. Jommi
Dipartimento di Ingegneria Strutturale, Politecnico di Milano, Italy

ABSTRACT: On basis of experimentally determined retention curves, a model is proposed for the description of water content changes inside the macropores of an unsaturated soil. Two aspects of soil behaviour are taken into account: the hysteresis of the water storage mechanism and its dependence on the void ratio. A description is provided of the set of retention curves, enveloping all the possible soil states in the void ratio-water ratio-suction space. This description enables to develop an elastoplastic model, which copes with the hysteretic effect in the retention curve and the onset of irreversible changes in water content during deformation. A hydro-mechanical model for unsaturated soils is then established by coupling the mentioned formulation and the Barcelona Basic Model (BBM). The formulation of the model is presented in the paper and illustrated on two examples.

1 INTRODUCTION

The relationship between suction and water content at equilibrium (the so-called *retention curve*) has been often disregarded in unsaturated soil mechanics, both in experimental investigations and in constitutive modelling. It was thought that the retention curve could be relevant only in modelling flow and transport phenomena within porous media. Thus, interest was shown in the role played by capillarity in seepage, especially with reference to earth dams problems, taking into account its influence on the hydraulic conductivity. Barbour (1998) has presented a historical review of studies on the topic.

It is however worth noting that in '40 Childs had already observed that the retention curve must be considered a complementary part to mechanical analysis, because it embodies valuable information on void size distribution of porous media. Nevertheless, for a long time, the role of negative pore water pressure has been taken into account only in relation to its contribution to the shear strength.

During the 70's, comprehensive frameworks for the behaviour of unsaturated soils began to be presented (see Wheeler & Karube 1996, for a review of constitutive modelling of unsaturated soils). Adopting the stress in excess to pore air pressure (i.e. the net mean stress p'' or the net vertical stress σ''_v) and the air pressure in excess to water pressure (i.e. the matric suction s) as controlling variables, state surfaces were initially proposed to describe the hydro-mechanical behaviour of an unsaturated soil. They relate the void ratio e and the degree of saturation S_r (or the water content w) to both p'' (or σ''_v) and s. The major drawback of this approach is the inability to represent irreversible behaviour.

The first model able to deal with irreversible effects in the response of unsaturated soils was formulated by Alonso et al. (1990) within the framework of elastoplasticity. Following that work, Wheeler & Sivakumar (1995) and Wheeler (1996) proposed an enhanced elastoplastic model, which takes into account variations of water content w. At the same time, relationships between the retention curve and other material variables, such as shear strength, permeability, and thermal conductivity have been proposed (Barbour 1998).

New advances in experimental techniques allow now validating the constitutive assumptions on the bases of reliable laboratory data. Accurate water volume change measurement has been achieved in experimental equipment (see e.g. Rampino et al. 1999), and laboratory investigations have been carried out to analyse the dependence of the retention curve on soil structure, stress level and stress history (see e.g. Vanapalli et al. 1999).

Barbour (1998), Romero (1999) and Vanapalli et al. (1999) pointed out that two main mechanisms generally govern the storage of water inside a soil. The first mechanism is mainly related to free water flow inside the macropores. The second one is related to water adsorption at the intra-aggregate level, when it exists. While the second mechanism is virtually independent of the macroscopic structure, the first one is coupled with the mechanical response of the soil. Increasing experimental evidence (Wheeler 1996; Rampino et al. 1999; Romero 1999) shows that irreversible changes in water content may occur as a consequence of volumetric deformation of the soil skeleton. On the other hand, drying and wetting history of the soil may affect the evolution of the irreversible strains (Chen et al. 1999).

In the following, a model is proposed to describe the two main characteristics of water storage mechanism in the macropores of a soil: the dependence of the retention curve on the void ratio and the hysteretic change in water content during drying-wetting cycles. Relating the change in void ratio to the mechanical response of the soil by the Barcelona Basic Model (BBM, Alonso et al. 1990), a coupled elastoplastic hydro-mechanical model is obtained. It provides a simple description of reversible and irreversible changes in strain and water content under stress and suction changes.

2 WATER STORAGE MECHANISMS IN A CLAYEY SOIL

The experimental observations and the theoretical considerations, on which the model is based, are presented in a companion paper (Romero & Vaunat 2000). They are briefly recalled in this section.

As already mentioned, the change in water content in a clayey soil is mainly related to variation of water volume in the macro- and in the inter-aggregate pores. The change in water content in the macropores is usually induced by experimental techniques causing variations in the gas or liquid pressures and consequently in the matric suction. The changes in water content at the intra-aggregate level are mainly induced by experimental techniques controlling the relative humidity. By use of the psychrometric law, the energy needed to reach a given water content under controlled relative humidity can be expressed in terms of a pressure, namely the total suction. For any water content, the water storage capacity of a clayey soil can thus be represented as a function of a unique pressure-dimension variable. The generic term *suction* will be adopted for this variable.

The relationship between water content and suction obtained on a kaolinitic-illitic clay (Boom clay, $w_L = 56\%$, $w_P = 29\%$) at two dry densities is depicted in Figure 1. The characterisation of the clay is given in Romero & Vaunat (2000). The low-density packing has a void ratio e_0 after compaction equal to 0.93. The high-density packing has a void ratio after compaction equal to $e_0 = 0.60$. During hydration, the volume of the sample is maintained constant. However, some small change in density can not be avoided (intrusion of material inside the porous stone during expansion, small collapse in case of the low-density packing, etc.). Therefore, the average values of void ratios after hydration are 0.90 and 0.65 for the respective packings.

In Figure 1, the zones of the two distinct storage mechanisms can be observed. For a water content greater than 15%, the water is free to move under hydraulic gradients (Romero et al. 1999). This zone corresponds to the zone of *free water*, present inside the macropores. Below 15%, the water content can be changed by controlling the relative humidity. This zone corresponds to the zone of *adsorbed water*, present inside the intra-aggregate pores and weakly bonded. A third zone can be detected below the residual water content, w_r (close in this case to 2%). It corresponds to the zone of *strongly bonded water* to the solid crystals.

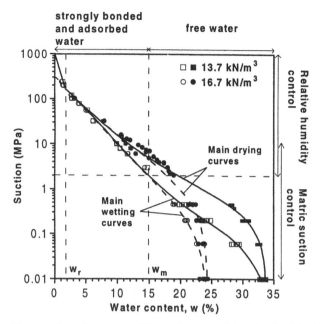

Figure 1. Retention curves of Boom clay at two initial dry densities (from Romero 1999).

The amount of strongly bonded and adsorbed water is mostly a function of the physical and chemical properties of the soil particles and of the pore fluids. It can be measured by the *microscopic water content* w_m, equal to the weight of strongly bonded and adsorbed water divided by the weight of solid. This quantity is independent of void ratio and depends mainly on the specific surface of the particles (Romero & Vaunat 2000). For Boom clay, it has a value of 15% (Romero 1999). At the microscopic water content, all the macropores are dry.

The amount of free water depends on the values of void ratio e and matric suction s. The most relevant aspects of the relationship between w, s and e are (see Figure 1):

1. all the possible states are located within a zone limited by the main drying curve MC_I and the main wetting curve MC_D,
2. cycles of suction may cause hysteretic changes in water content,
3. soil deformation under mechanical action causes changes in water content.

These three aspects are discussed in the following paragraphs.

2.1 *Main drying and main wetting curves at constant void ratio*

The retention curve depicted in Figure 1 for the dry density of 13.7 kN/m^3 is first considered.

An increase in suction from the saturated state ($w = 35\%$) will first cause a slight decrease in water content. When the suction exceeds the air entry value (estimated between 0.1 and 0.2 MPa), a sharp reduction in water content is observed. As drying proceeds, this reduction is more pronounced. Throughout this process, the hydraulic state of the soil follows the main drying curve MC_I. At w_m, a change in behaviour is observed. In fact, further increase in suction affects the adsorbed water, until w_r is reached.

A decrease in suction from the value corresponding to the dry state ($w_r = 2\%$) causes an increase in water content along a path different from that followed during drying. This path is described by the main wetting curve MC_D. As long as the water content remains below the value of the microscopic water content, only a change of water content in the intra-aggregate pores is induced. Above the microscopic water content, the decrease in suction increases the content in free water. The water content at null suction may be slightly less than the water content of the initially saturated sample, because of the presence of occluded air. For the sake of simplicity, the latter effect will not be taken into account in the formulation, and the main wetting curve will be considered joining the main drying curve at the same water content for s = 0.

2.2 *Hysteretic character of the retention curve*

If the drying process from the saturated state is stopped before the residual water content is reached and the sample is wetted, a hysteretic behaviour is observed. The increase of water content at the beginning of the wetting process is small and the state of the soil follows a curve, usually referred to as *scanning curve*, having a slope steeper than the slopes of the main curves. When the state of the sample reaches the intersection between the scanning curve SC and the main wetting curve MC_D, it stops following SC and starts to move along MC_D. The main wetting curve acts then as a state boundary line, which divides the w-s plane into the zone of attainable states and the zone of unattainable states during wetting at a given void ratio. If the increment of suction is reverted, the state of the soil will follow a new scanning curve until it reaches the main drying curve MC_I. MC_I can then be also seen as a state boundary line, separating the zone of attainable states from the zone of unattainable states, during drying at fixed e. Consequently, at a given void ratio, none of the states above the main drying curve and below the main wetting curve can be reached by changing suction, while all the states between the two curves can be obtained by an appropriate drying-wetting cycle starting from saturated conditions.

The distinct paths involved during a drying-wetting cycle evidence irreversible changes in water content along the main curves. Nitao & Bear (1996) invoke different mechanisms for this non-reversible behaviour. At water contents below w_m, the irreversibility is related to hysteresis in chemical potentials. At water contents above w_m, hysteresis is mostly governed by the instability of the interfaces between the gas and the liquid phases inside the macropores. In this case, hysteresis can be related to the dissipative process involved during abrupt switching of the interfaces from unstable to more stable geometry during drying or wetting.

Hysteresis can be also observed, in a less pronounced way, during cycles of suction inside the zone delimited by the two main curves. This fact indicates that some irreversible changes in water content may occur along the scanning curve. The irreversibility generally increases as the state of the soil approaches the main curves. This phenomenon is not considered in the model and the hydraulic response along any SC is assumed reversible.

2.3 *Evolution of the retention curve with void ratio*

The comparison between the retention curves obtained for the two distinct packings provides some insights into its dependence on void ratio.

In the range of water contents below w_m, the two main curves are superimposed, suggesting that the macrostructure built during the compaction does not have any effect on the water distribution inside the micropores. This has been discussed by Vanapalli et al. (1999), who note that the dependence of the retention curve on the stress level may be noticeable in soils possessing a definite macrostructure, while it is negligible for soils that present a matrix-dominant behaviour. Similar conclusions were drawn by Romero et al. (1999).

In the range of water contents above w_m, the retention curve for the two packings markedly diverge in order to reach the saturated water contents corresponding to the final void ratios at null suction. Moreover, the shape of the main curves appears to be different between the two packings. As discussed by Romero & Vaunat (2000), this is mainly due to the fact that the air entry value is affected by the value of void ratio. At the present stage of the model development, this latter effect is disregarded and, in a first approximation, the retention curve is assumed to keep the same shape at distinct void ratios. Since the water ratio must be equal to the void ratio at null suction, the retention curve has to be expressed as a homothetical function of e.

A consequence of the dependence of the retention curve on void ratio is that the storage capacity in the macropores of the soil is affected by stress history. Thus, loading or unloading at constant suction in the zone of free water (w $\geq w_m$) may cause reversible and irreversible changes in water content. On the other hand, it is known that an increase or a decrease in suction at constant net stress may induce either elastic or elastoplastic volumetric strain. The model presented in this paper aims at representing this symmetric hydro-mechanical behaviour.

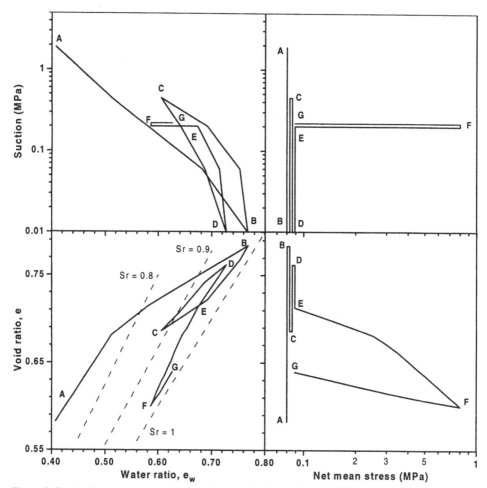

Figure 2. Results from isotropic test in the $p"$-s-e_w-e plot (from Romero 1999).

3 MODEL CONCEPTUAL BACKGROUND

3.1 *Work-conjugate variables*

Work-conjugate variables should be preferably adopted in developing models. They are defined in such a way that the rate of input work per unit volume of medium equates to the sum of the products of all stress components multiplied by their corresponding components in strain rate.

In an unsaturated soil, the total stresses σ_{ij}, the pressure of the gas phase u_a and the pressure of the liquid phase u_w generally define the stress state. Nevertheless, experimental results from null-type tests are giving increasing evidence that any pair of combinations between σ_{ij}, u_a and u_w may be adopted to describe the behaviour of the soil. For experimental convenience, the net stress $\sigma"_{ij} = \sigma_{ij} - u_a \delta_{ij}$ and the suction $s = u_a - u_w$ are usually adopted.

Edgar (1993), Wheeler & Sivakumar (1996), Houlsby (1997) and Dangla et al. (1997) converge to the same statement concerning the variables work-conjugated to the net stresses and the suction. Neglecting the contributions of fluids flow and air compressibility, Houlsby (1997) expresses the rate of input work per unit volume $\dot{W}$ as:

$$\dot{W} = \sigma"_{ij}\dot{\varepsilon}_{ij} + s\dot{\varepsilon}_w \qquad \text{(Soil Mechanics convention is used)} \qquad (1)$$

125

in which ε_{ij} is the usual strain tensor and ε_w is a generalised strain component associated to suction, called hydraulic strain. ε_w is equal to $- e_w / (1 + e_0)$, where e_0 is the void ratio of the reference configuration and e_w is the water ratio (water volume divided by solid volume). With this definition, $\dot{\varepsilon}_w$ is equal and opposite to the increment of volumetric water content referred to the reference configuration.

3.2 Representation of isotropic hydro-mechanical states

Under isotropic conditions, equation (1) becomes:

$$\dot{W} = p''\dot{\varepsilon}_v + s\dot{\varepsilon}_w \tag{2}$$

where p'' is the net mean stress, $\dot{\varepsilon}_v = -\dot{e}/(1+e_0)$ is the volumetric strain and $\dot{e}$ is the void ratio increment.

The coupled changes in net mean stress, suction, volumetric strain and volumetric water content are totally described by a curve in the p''-e-s-e_w space. The projection of this curve on the four planes p''-e, p''-s, e_w-s and e_w-e gives a full representation of the hydro-mechanical response of the soil.

The e-p'' and p''-s planes are typically used to describe the mechanical response of unsaturated soil under isotropic stress and suction changes.

The e_w-s plane is adopted to represent the retention curve. In fact, multiplying the water content by the specific gravity G_s, the water ratio is recovered. In this plane, the retention curve ends at the void ratio at null suction. Moreover, the water ratio remains constant during an undrained test, as the void ratio does during an isochoric test. This provides a useful symmetry between the mechanical plane e-p'' and the hydraulic plane e_w-s.

The plane of the kinematical variables e_w-e enables to follow the evolution of the degree of saturation. It is equivalent to the dry unit weight γ_d – gravimetric water content w plane, in which the results of compaction tests are usually depicted.

An example of the representation of wetting-drying and loading-unloading paths in the four planes is given in Figure 2.

3.3 Modelling of the water retention curves

Since the present work focuses on the coupled hydro-mechanical response of unsaturated materials, the formulation will be restricted to the range of water ratios affected by mechanical actions, that is between $e_w = e_{wm} = w_m G_s$ and $e_w = e$. e_{wm} is called water ratio cut-off. In this range, the main wetting and the main drying curves are modelled through a van Genuchten' s expression modified by Romero (1999).

The expression for the main drying curve is given by:

$$\frac{e_{wI} - e_{wm}}{e - e_{wm}} = \Psi_I(s) = \left(1 + (\alpha_I s)^n\right)^{-m}\left(1 - \frac{\ln\left(1 + \dfrac{s}{\beta_I}\right)}{\ln(2)}\right) \tag{3}$$

where $1/\alpha_I$ is related to the air-entry value of the packing, n is related to the slope of the curve for suctions higher than the air-entry value, m is related to the residual water ratio and β_I is the suction corresponding to the water ratio e_{wm}.

Similarly, the main wetting curve is expressed by:

$$\frac{e_{wD} - e_{wm}}{e - e_{wm}} = \Psi_D(s) = \left(1 + (\alpha_D s)^n\right)^{-m}\left(1 - \frac{\ln\left(1 + \dfrac{s}{\beta_D}\right)}{\ln(2)}\right) \tag{4}$$

Equations (3) and (4) give an automatic scaling of the main curves with respect to the void ratio, according to the concepts presented in paragraph 2.3.

Table 1. Material parameters of the retention curve for Boom clay.

Parameter	e_{wm}	n	m	α_I (MPa^{-1})	β_I (MPa)	α_D (MPa^{-1})	β_D (MPa)
Value	0.4	0.4	0.8	0.4	5	1.5	2

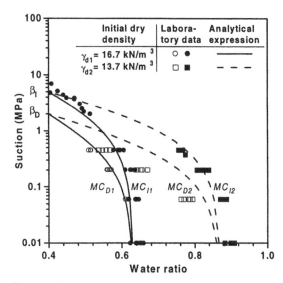

Figure 3. Retention curve for Boom clay: analytical expression *vs.* experimental data.

Parameters adopted for Boom clay are summarised in Table 1. In Figure 3, the corresponding analytical expressions are compared with experimental data. For the highest density, the retention curve fits well the experimental data. For the lowest density, the main drying curve is well represented by the analytical function. A noticeable discrepancy between the analytical expression and laboratory data can be observed for the main wetting curve, due to the fact that, at present, the model considers the air-entry pressure value independent of the void ratio. For a more accurate modelling of soil behaviour, relationships between α_D, α_I and e have to be considered.

4 MODEL FORMULATION

4.1 *Mechanical behaviour*

The mechanical behaviour is based on the Barcelona Basic Model (BBM) developed by Alonso et al. (1990). The bases of the model are summarised hereafter, considering the case of isotropic stress conditions only.

Consider an initial stress state M located below the isotropic consolidation line *ICL* of the material (see Figure 4). On a loading path from M at a constant suction s, a slight decrease in void ratio is first observed, linearly varying with the logarithm of the mean net stress, and recoverable upon unloading (path MM$_1$ in Figure 4). Similarly, a reversible decrease in void ratio is recorded on drying from M, linearly dependent on the logarithm of suction (path MM$_4$ in Figure 4). When suction and mean net stress are simultaneously changed, both effects contribute to the reversible variation of void ratio. This behaviour is interpreted as an elastic response of the soil skeleton. Assuming small strains, the elastic law is expressed by:

$$d\varepsilon_v^e = \frac{dp''}{K_t} + \frac{ds}{K_s} \quad \text{with} \quad K_t = \frac{(1+e)p''}{\kappa} \quad \text{and} \quad K_s = \frac{(1+e)(s+p_{at})}{\kappa_s} \tag{5}$$

where $d\varepsilon_v^e$ is the increment of volumetric elastic strain, K_t is the mechanical stiffness against stress changes, K_s is the mechanical stiffness against suction changes, κ is the slope of the unloading-reloading line in the e-$ln(p'')$ diagram at constant s, κ_s is the slope of the wetting-drying line in the e-$ln(s+p_{at})$ diagram at constant p'', and p_{at} is the atmospheric pressure.

On further loading at a constant suction s, a sudden decrease in void ratio is recorded when p'' passes the threshold p_0. This change in behaviour is explained by development of plastic strain. The p_0 value is interpreted as the yield point, intersection of a yield surface formulated in the space of the six stress components, with the isotropic axis. When only isotropic compression stress conditions are considered, a simplified expression for the yield surface is:

$$LC = p'' - p_0 = 0 \tag{6}$$

BBM generally defines the direction of plastic strain increments by a non-associated flow rule, chosen in order to allow for null lateral strain on oedometer loading. For isotropic compression stress conditions, the expression of the flow rule reduces to:

$$d\varepsilon_v^p = r_v^{LC} d\gamma^{LC} \qquad \text{with} \qquad r_v^{LC} = 1 \tag{7}$$

where $d\varepsilon_v^p$ is the volumetric plastic strain increment, $d\gamma^{LC}$ is the plastic multiplier associated to LC and r_v^{LC} is the flow rule for isotropic stress conditions.

From p_0 on (path M_1M_2 in Figure 4), the soil follows the isotropic compression line ICL corresponding to the suction s. If the sample is unloaded from point M_2 to point M_3 and reloaded, the soil behaves elastically until p'' exceeds again p_0. This means that all the points below ICL can be reached following an appropriate loading-unloading path, while none of the points above this line can be reached by isotropically changing the stresses. In other words, ICL is a state boundary line in the e-$ln(p'')$ diagram, which separates the attainable states from the unattainable states for the given suction s. The restriction that the state of the soil must remain on ICL during plastic loading enables the hardening law to be formulated at a constant suction:

$$d\varepsilon_v^p = d\varepsilon_v - d\varepsilon_v^e = -\frac{de}{1+e} - \frac{\kappa}{1+e}\frac{dp_0}{p_0} = \frac{\lambda(s) - \kappa}{1+e}\frac{dp_0}{p_0} \tag{8}$$

where $d\varepsilon_v$ is the volumetric strain increment and $\lambda(s)$ is the slope of ICL. The dependence of λ on s is expressed by default through the relationship $\lambda(s) = \lambda(0)[r + (1-r)e^{-\beta s}]$, where $\lambda(0)$ is the slope of the isotropic compression line in saturated conditions and r and β are two material parameters.

If the sample is dried from point M to a suction s^* (point M_4 in Figure 4) and then loaded, the yield point $p_0^*(s^*)$ (point M_5 in Figure 4) will be higher than $p_0(s)$. Moreover, a decrease of suction back to s at a constant mean net stress brings the void ratio from point M_5 to point M_2 on the isotropic consolidation line. This indicates that the yield point reached from a given initial void ratio can be related uniquely to the final value of suction and mean net stress, whatever stress-suction path followed is. Equating the value of the void ratio at point M_4, the following relationship between p_0 and p_0^* can be deduced:

$$(\lambda(s) - \kappa)\ln\left(\frac{p_0}{p_c}\right) + \kappa_s \ln(s + p_{at}) - e_N(s) = (\lambda^* - \kappa)\ln\left(\frac{p_0^*}{p_c}\right) + \kappa_s \ln(s^* + p_{at}) - e_N^* \tag{9}$$

where p_c is a reference pressure, $e_N(s)$ is the intercept of ICL with the line $p = p_c$, $\lambda^* = \lambda(s^*)$ is the slope of ICL^* and $e_N^* = e_N(s^*)$ is the intercept of ICL^* with the line $p = p_c$.

Equation (9) relates implicitly the yield point at M_1 to the yield point at M_5 along the elastic plane $MM_1M_5M_4$. The set of solutions (9) can be geometrically represented by the intersection of the elastic plane passing through M with the surface formed by all the isotropic compression lines. This surface forms a stable state boundary surface that splits the e-$ln(p'')$-$ln(s+p_{at})$ space into the domains of possible and impossible states (see Figure 5). The intersection previously mentioned gives the shape of the yield function LC in the net mean stress-suction plane.

Equation (9) can be simplified by considering the additional assumption that, at the reference pressure p_c, all the $e_N(s)$ are located along the elastic line in the e-$ln(s+p_{at})$ diagram, that can be expressed as follows:

$$e_N(s) = e_N(0) - \kappa_s \ln\left(\frac{s + p_{at}}{p_{at}}\right) \tag{10}$$

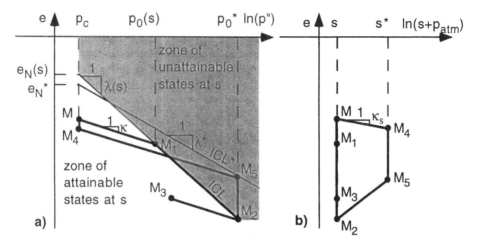

Figure 4. Mechanical part of the model: a) soil response in the mean net stress – void ratio plane; b) soil response in the suction – void ratio plane.

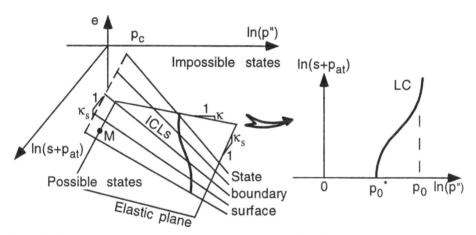

Figure 5. Mechanical part of the model: shape of yield function LC is the net mean stress – suction plane.

Using equation (10) and taking the saturated state ($s = 0$) as the reference state, the relation (9) turns into an explicit expression of LC curve:

$$(\lambda(s) - \kappa)\ln\left(\frac{p_0}{p_c}\right) = (\lambda(0) - \kappa)\ln\left(\frac{p_0^*}{p_c}\right) \qquad \text{or} \qquad \left(\frac{p_0}{p_c}\right) = \left(\frac{p_0^*}{p_c}\right)^{\frac{\lambda(0) - \kappa}{\lambda(s) - \kappa}} \qquad (11)$$

Using equation (11), the hardening law (8) becomes:

$$dp_0^* = \frac{(1+e)p_0^*}{\lambda(0) - \kappa} d\varepsilon_v^p = \frac{(1+e)p_0^*}{\lambda(0) - \kappa} r_v^{LC} d\gamma^{LC} = H^{LC} d\gamma^{LC} \qquad (12)$$

Equation (5), equations (6) plus (11), equation (7) and equation (12) provide the full set of mechanical elastoplastic constitutive equations (elastic law, yield surface, flow rule and hardening law) for isotropic stress conditions.

129

4.2 Hydraulic behaviour

The hydraulic response of the sample, that is the change in water content under mean net stress and suction increments, is now considered As discussed in section 3, this response is described in the $e_w - ln(s)$ plane. In this plane, the main wetting curve is characterised by:

$$e_{wD} = MC_D(e,s) = e_{wm} + (e - e_{wm}) \psi_D(s) \tag{13}$$

and the main drying curve by:

$$e_{wI} = MC_I(e,s) = e_{wm} + (e - e_{wm}) \psi_I(s) \tag{14}$$

where e_{wm} is the water ratio cut-off defined in paragraph 2 and ψ_D and ψ_I are expressed by equations (4) and (3) respectively. As already mentioned in section 3, MC_D and MC_I delimit two zones in the $e_w - ln(s)$ diagram. The zone between the main wetting curve and the main drying curve contains all the attainable states. Above the main drying curve and below the main wetting curve are located all the hydraulic states unattainable for a given void ratio e.

Let us consider an initial state H, located between the two main curves of the retention curve. If the suction is slightly changed, the water ratio moves along the scanning curve passing through H (path HH_1 in Figure 6). On the other side, if the sample is loaded, a small amount of water is expelled from the sample (path HH_4 in Figure 6). These changes in water ratio are considered reversible, as mentioned in section 3. Assuming small strains and noting $d\varepsilon_w = - de_w / (1 + e)$ the increment of the generalised strain associated with the suction, the hydraulic behaviour around H can be formulated by the following elastic law:

$$d\varepsilon_w^e = \frac{dp''}{K_{wt}} + \frac{ds}{K_w} \quad \text{with} \quad K_{wt} = \frac{(1+e)p''}{\kappa_{wt}} \quad \text{and} \quad K_w = \frac{(1+e)}{\kappa_w} \tag{15}$$

where superscript e stands for elastic, K_{wt} is the hydraulic stiffness against stress changes, K_w is the hydraulic stiffness against suction changes, κ_{wt} is the slope of the unloading-reloading line in the $e_w - ln(p'')$ diagram at constant s and κ_w is the slope of the scanning curve in the $e_w - s$ diagram at constant p''. Experimental evidence presented by Romero (1999) (see also Romero & Vaunat 2000) indicates that, for Boom clay, κ_w can be considered constant in the nearly saturated range. In case of another material or for a distinct range of suction, other expressions for K_w could be introduced without transgressing the framework of the present model.

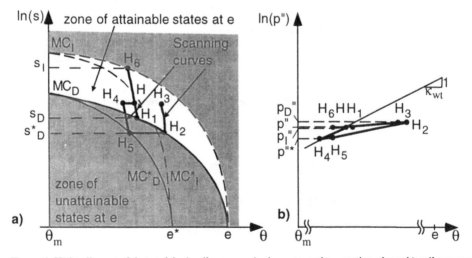

Figure 6. Hydraulic part of the model: a) soil response in the water ratio – suction plane; b) soil response in the water ratio – net mean stress plane.

If the sample is submitted to prolonged wetting at a constant void ratio, the main wetting curve MC_D will be reached at point H_1 (see Figure 6). The water ratio then follows MC_D (path H_1H_2 in Figure 6). When the sample is dried from point H_2, e_w moves back along a scanning curve (path H_2H_3 in Figure 6), indicating that the change in water ratio that occurred between H_1 and H_2 has an irrecoverable component. When re-wetted from H_3, the sample reaches again the main wetting curve at point H_2. The value of the suction at H_2 can then be interpreted as a yield point, labelled s_D, which determines the onset of plastic changes in water ratio on wetting. The corresponding yield function SD is:

$$SD = s_D - s = 0 \tag{16}$$

Since the main wetting curve depends on the value of void ratio e, s_D will also depend on e. On a drying path starting from H, the behaviour is totally symmetric. A yield point s_I, depending on the void ratio, can be associated to the point where the scanning curve passing through H intersects the main drying curve MC_I (path HH_6 in Figure 6). The corresponding yield function SI is:

$$SI = s - s_I = 0 \tag{17}$$

where s_I is a function of the void ratio. The yield points s_D and s_I are related to each other. In fact, starting from H and changing the suction in such a way that only elastic changes in water ratio occur, both yield points can be reached. Noting that an increment of suction in the elastic domain at a constant void ratio e causes a slight change in the mean net stress, the condition relating s_D and s_I may be written:

$$e_{wD}(s_D) - e_{wI}(s_I) = -\kappa_w(s_D - s_I) - \kappa_{wt}(\ln(p''_D) - \ln(p''_I)) \tag{18}$$

where p''_D and p''_I are the mean net stress at H_1 and H_6, respectively. The condition that the volumetric elastic strain is equal to zero between H and H_1 leads to the following relationship between p''_D and s_D:

$$\Delta\varepsilon_v^e\Big|_H^{H_1} = \frac{\kappa}{1+e}\left(\ln(p''_D) - \ln(p'')\right) + \frac{\kappa_s}{1+e}\left(\ln(s_D + p_{at}) - \ln(s + p_{at})\right) = 0 \tag{19}$$

In the same way:

$$\Delta\varepsilon_v^e\Big|_H^{H_6} = \frac{\kappa}{1+e}\left(\ln(p''_I) - \ln(p'')\right) + \frac{\kappa_s}{1+e}\left(\ln(s_I + p_{at}) - \ln(s + p_{at})\right) = 0 \tag{20}$$

and equation (18) becomes:

$$e_{wD}(e, s_D) + \kappa_w s_D - \frac{\kappa_{wt}\,\kappa_s}{\kappa}\ln(s_D) = e_{wI}(e, s_I) + \kappa_w s_I - \frac{\kappa_{wt}\,\kappa_s}{\kappa}\ln(s_I) \tag{21}$$

The flow rules of SD and SI have the trivial expressions:

$$r^{SD} = 1 \quad \text{and} \quad r^{SI} = -1 \tag{22}$$

As in the case of the mechanical model, the hardening laws are deduced from the fact that the main curves are state boundary lines in the e_w - s diagram at a constant void ratio. Along MC_D, the change in water ratio at a constant void ratio is expressed as:

$$de_w = \frac{\partial MC_D}{\partial s_D}ds_D = -\lambda_{wD}(e_w)\,ds_D \tag{23}$$

where λ_{wD} is the opposite of the slope of the main wetting curve at e_w. The restriction that e_w must remain on MC_D during a plastic wetting implies that:

$$d\varepsilon_w^p = d\varepsilon_w - d\varepsilon_w^e = -\frac{de_w}{1+e} - \frac{\kappa_w}{1+e}ds_D - \frac{dp''}{K_{wt}} = \frac{\lambda_{wD} - \kappa_w}{1+e}ds_D - \frac{dp''}{K_{wt}} = r^{SD}d\gamma^{SD} \tag{24}$$

where $d\gamma^{SD}$ is the plastic multiplier associated with SD yield surface. At a constant void ratio, dp'' is related to the increment of the plastic multiplier $d\gamma^{LC}$ by:

$$d\varepsilon_v = d\varepsilon_v^p + d\varepsilon_v^e = r_v^{LC} d\gamma^{LC} + \frac{dp''}{K_t} + \frac{ds_D}{K_s} = 0 \tag{25}$$

that is:

$$dp'' = \frac{p''(1+e)}{\kappa} r_v^{LC} d\gamma^{LC} - \frac{\kappa_s}{\kappa} \frac{p''}{s + p_{at}} ds_D \tag{26}$$

Equations (24) and (26) lead to the following hardening law:

$$ds_D = \frac{1+e}{\lambda_{wD} - \kappa_w + \dfrac{\kappa_{wt} \kappa_s}{\kappa(s_D + p_{at})}} \left(r^{SD} d\gamma^{SD} - \frac{\kappa_{wt}}{\kappa_w} r_v^{LC} d\gamma^{LC} \right) \tag{27}$$

The term $\kappa_w - (\kappa_{wt}\,\kappa_s)/(\kappa\,(s_D + p_{at}))$ takes into account the elastic change in e_w during hardening at a constant void ratio and is noted $\overline{\kappa_{wD}}$. Equation (27) introduces a coupling between the hardening of LC and SD yield surfaces. If no plastic deformation takes place ($d\gamma^{LC} = 0$), the hardening of SD is only caused by plastic change in water ratio. The hardening law of s_I during drying has the symmetric expression:

$$ds_I = \frac{1+e}{\lambda_{wI} - \overline{\kappa_{wI}}} \left(r^{SI} d\gamma^{SI} - \frac{\kappa_{wt}}{\kappa_w} r_v^{LC} d\gamma^{LC} \right) \tag{28}$$

where $d\gamma^{SI}$ is the plastic multiplier associated with SI yield surface and κ_{wI} is equal to $\kappa_w - (\kappa_{wt}\,\kappa_s)/(\kappa\,(s_I + p_{at}))$. Due to the fact that s_D and s_I are related to each other by equation (21), hardening of s_D occurs also during plastic drying. Differentiation of (21) at constant void ratio leads to:

$$\left(\lambda_{wD} - \overline{\kappa_{wD}}\right) ds_D = \left(\lambda_{wI} - \overline{\kappa_{wI}}\right) ds_I \tag{29}$$

Substitution of equation (29) into (28) provides the expression for the hardening of SD when SI is active:

$$ds_D = \frac{1+e}{\lambda_{wD} - \overline{\kappa_{wD}}} \left(r^{SI} d\gamma^{SI} - \frac{\kappa_{wt}}{\kappa_w} r_v^{LC} d\gamma^{LC} \right) \tag{30}$$

In the same way, the hardening of SI when SD is active is deduced from equations (29) and (27):

$$ds_I = \frac{1+e}{\lambda_{wI} - \overline{\kappa_{wI}}} \left(r^{SD} d\gamma^{SD} - \frac{\kappa_{wt}}{\kappa_w} r_v^{LC} d\gamma^{LC} \right) \tag{31}$$

If the sample is loaded elastically from point H at a mean net stress $p''*$ to reach a void ratio $e*$ smaller than e (path HH$_4$ in Figure 6) and then wetted, a new main wetting curve MC$*_D$ corresponding to $e*$ is reached (point H$_5$ in Figure 6). As the main wetting curve acts as a state boundary line, unloading from H$_5$ back to the initial value of the void ratio e will cause an increase in the water ratio such that the state of the soil remains on MC_D. Moreover, the uniqueness of the relationship between MC_D and the value of void ratio (equation (4)) imposes that the state of the sample will end at H$_2$. The relationship between s_D and $s*_D$ can thus be deduced by equating the value of water ratio at H$_4$:

$$MC_D(e, s_D) + \kappa_w s_D + \kappa_{wt} \ln(p'') = MC_D^* + \kappa_w s_D^* + \kappa_{wt} \ln(p''*) \tag{32}$$

with $e = e* - \kappa\,(ln(p'') - ln(p''*)) - \kappa_s\,(ln(s_D + p_{at}) - ln(s*_D + p_{at}))$. Equation (32) is the hydraulic counterpart of (9). It expresses implicitly the relationship between the hydraulic yield point at H$_1$ and the hydraulic yield point at H$_5$ along the elastic plane HH$_1$H$_5$H$_4$. It is the mathematical expression of the intersection in the space e-e_w-$ln(s+p_{at})$ of the boundary state surface enveloping the main wetting curves for all possible void ratios with the elastic plane passing through H. This intersection provides the shape of the yield surface in the $ln(p'')$-$ln(s+p_{at})$ plane shown in Figure 7. It is generally impossible to derive explicitly the relationship between s_D and $s*_D$ from

132

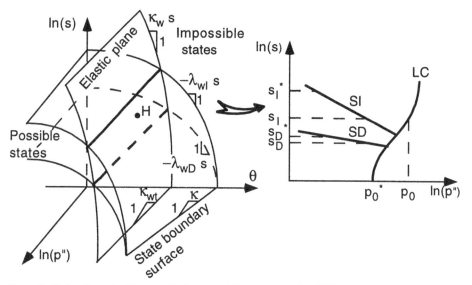

Figure 7. Hydraulic part of the model: shape of yield functions SI and SD in the mean net stress – suction plane.

equation (32). By its differentiation at constant void ratio, it is however possible to compute ds_D as a function of $ds*_D$:

$$\left(\lambda_{wD} - \overline{\kappa_{wD}}\right) ds_D = \left(\lambda^*_{wD} - \kappa^*_{wD}\right) ds^*_D \tag{33}$$

where λ^*_{wD} is equal to $-\partial MC^*_D / \partial s^*_D$ and $\overline{\kappa^*_{wD}}$ is equal to $\kappa_w - (\kappa_{wt} \kappa_s) / (\kappa (s^*_D + p_{at}))$. Similarly, s_I and $s*_I$ are related by:

$$MC_I(e, s_I) + \kappa_w s_I + \kappa_{wt} \ln(p'') = MC^*_I + \kappa_w s^*_I + \kappa_{wt} \ln(p''*) \tag{34}$$

with λ^*_{wI} equal to $-\partial MC^*_I / \partial s^*_I$ and $\overline{\kappa^*_{wI}}$ equal to $\kappa_w - (\kappa_{wt} \kappa_s) / (\kappa (s^*_I + p_{at}))$. The following relationship between ds_I and $ds*_I$ is then obtained:

$$\left(\lambda_{wI} - \overline{\kappa_{wI}}\right) ds_I = \left(\lambda^*_{wI} - \overline{\kappa^*_{wI}}\right) ds^*_I \tag{35}$$

Choosing $s*_D$ and $s*_I$ as the hardening parameters of SD and SI yield surfaces, the hardening laws can be summarised synthetically by the expression:

$$ds^*_\alpha = \frac{1+e}{\lambda^*_{w\alpha} - \kappa^*_{w\alpha}} \left(-\frac{\kappa_{wt}}{\kappa_w} r^{LC}_v d\gamma^{LC} + r^{SD} d\gamma^{SD} + r^{SI} d\gamma^{SI} \right) \qquad \alpha = D, I \tag{36}$$

$$= H^{LC}_\alpha d\gamma^{LC} + H^{SD}_\alpha d\gamma^{SD} + H^{SI}_\alpha d\gamma^{SI}$$

Equation (17), equations (18) plus (32), equations (19) plus (34) and equation (36) provide the full set of hydraulic elastoplastic constitutive equations (elastic law, yield surface, flow rule and hardening law) of the model.

4.3 Coupled hydro-mechanical behaviour

The set of equations presented before describes the coupled hydro-mechanical behaviour of the soil. They are summarised in Table 2.

133

Table 2. Summary of elastoplastic equations of the coupled hydro-mechanical model.

	Mechanical behaviour	Hydraulic behaviour
Stress variable	$p''= p - u_a$	$s = u_a - u_w$
Kinematical variable	$e = V_V/V_S$	$e_w = V_W/V_S$
Strain variable	$\varepsilon_v = -e/(1+e_0)$	$\varepsilon_w = -e_w/(1+e_0)$
Hardening variable	p_0^*	s_D^*, s_I^*
Plastic multiplier	$d\gamma^{LC}$	$d\gamma^{SD}, d\gamma^{SI}$
Elastic law (small strains are assumed)	$d\varepsilon_v^e = \dfrac{dp''}{K_t} + \dfrac{ds}{K_s}$ with $K_t = \dfrac{(1+e)p''}{\kappa}$ and $K_s = \dfrac{(1+e)(s+p_{at})}{\kappa_s}$	$d\varepsilon_w^e = \dfrac{dp''}{K_{wt}} + \dfrac{ds}{K_w}$ with $K_{wt} = \dfrac{(1+e)p''}{\kappa_{wt}}$ and $K_w = \dfrac{(1+e)}{\kappa_w}$
Yield surface	$p''-p_0(p_0^*,s)=0$ with $p_0(p_0^*,s)$ given by (9)	$s_D(s_D^*,e) - s = 0$ $s - s_I(s_I^*,e) = 0$ with $s_D(s_D^*,e)$ given by (32) and $s_I(s_I^*,e)$ given by (34)
Flow rule	$d\varepsilon_v^p = r_v^{LC}d\gamma^{LC}$ with $r_v^{LC} = 1$	$d\varepsilon_w^p = r^{SD}d\gamma^{SD} + r^{SI}d\gamma^{SI}$ with $r^{SD} = -1$, $r^{SI} = 1$
Hardening law	$dp_0^* = H^{LC}d\gamma^{LC}$ with $H^{LC} = \dfrac{(1+e)p_0^*}{\lambda(0)-\kappa} r_v^{LC}$	$ds_\alpha^* = H_\alpha^{LC}d\gamma^{LC} + H_\alpha^{SD}d\gamma^{SD} H_\alpha^{SI}d\gamma^{SI}$ with $H_\alpha^{LC} = -\dfrac{1+e}{\lambda_{w\alpha}^* - \kappa_{w\alpha}^*}\dfrac{\kappa_{wt}}{\kappa_w} r_v^{LC}$, $H_\alpha^{SD} = \dfrac{1+e}{\lambda_{w\alpha}^* - \kappa_{w\alpha}^*} r^{SD}$, $H_\alpha^{SI} = \dfrac{1+e}{\lambda_{w\alpha}^* - \kappa_{w\alpha}^*} r^{SI}$ and $\lambda_\alpha^* = -\dfrac{\partial MP_\alpha^*}{\partial s}\Big|_{s_\alpha^*}$, $\overline{\kappa_{w\alpha}} = \kappa_w - \dfrac{\kappa_{wt}\kappa_s}{\kappa(s_\alpha^*+p_{at})}$ for $\alpha = D,I$

Table 3. Values of the mechanical parameters used in the model.

Parameter	κ	$\lambda(0)$	κ_s	p_c (MPa^{-1})	r	β (MPa^{-1})	κ_{wt}
Value	0.015	0.108	0.0012	6.547 10^{-3}	0.911	0.00575	1. 10^{-6}

5 ILLUSTRATIVE EXAMPLES

The original aspects of the model are illustrated on two examples. They aim at reproducing changes in water content measured in the laboratory on Boom clay samples. The hydraulic parameters used in the simulation are given in Table 1 (paragraph 3.3). The values of the mechanical parameters are given in Table 3. The initial values of the hardening parameters $p*_0$, $s*_l$ and $s*_D$ are taken equal to 10 MPa, 5 MPa and 1.7 MPa, respectively. The reference void ratio $e*$ is equal to 0.63.

The first example deals with suction cycles at a constant void ratio. The suction path imposed in the computation is depicted in Figure 8. In the same figure, the variations of water ratio given by the model are compared with data of the retention curve. Between the initial state A ($s = 5$ MPa and $e_w = 0.4$) and state B ($s = 0.5$ MPa), an elastic variation in water ratio is first observed. Once the yield point s_D is reached at A', irreversible changes in water ratio start. Between A' and B, SD yield surface is activated. In this interval, s_D is equal to the value of the imposed suction and the sample follows the main wetting curve. At B, a suction cycle between 0.5 MPa and 1 MPa is performed (path B-C-B). During this cycle, the state of the sample remains in the elastic zone between SD and SI and the variation of water ratio is reversible. At the end of the cycle, SD yield surface is reached again at $s = s_D$. The sample is afterwards brought close to saturation ($s = 0.01$ MPa, point D). This process is elastoplastic as evidenced by the evolution of yield point s_D. When the suction variation is reverted at D, an initial elastic response can be observed between D and D'. At D', SI yield surface begins to be activated. The e_w-s path follows then the main drying curve and s_I evolves in accordance with the value of the imposed suction.
During a moderate wetting-drying cycle (path E-F-E), the response of the soil is elastic. Example 1 shows how the hysteresis in the retention curve is captured through the proposed elastoplastic framework.

The second example handles the case of a wetting-loading process. The path followed in the p''-s diagram and the response of the sample are depicted in Figure 9. The sample is initially wetted from point A ($s = 5$ MPa and $e_w = 0.4$) to point B at a void ratio equal to 0.87. In this interval, the response given by the model is similar to that described in example 1. The evolution of the yield point s_I during activation of SD yield surface is moreover represented. Due to the coupling between SD and SI, s_I moves along the main drying curve during SD activation. At point B, mean net stress is increased in order to reach a final void ratio equal to 0.63. Due to the very small value considered for the coupling term κ_{wt}, the water ratio does not experience noticeable change during the loading stage. On the other hand, s_I decreases due to the decrease in void ratio. When s_I becomes equal to the current value of suction, the current state of the sample is reached by the main drying curve (point B'). Afterwards, due to the constraint that the zone

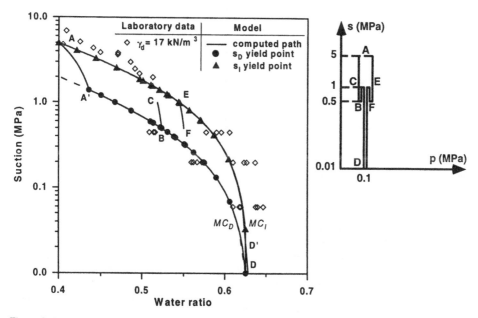

Figure 8. Example 1: simulation of a wetting-drying process at constant e.

135

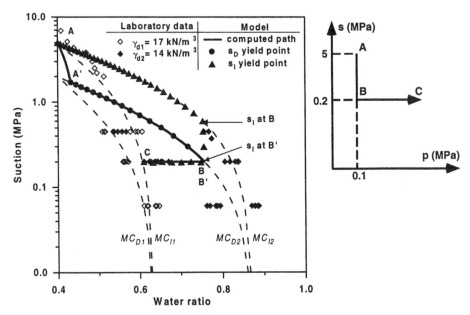

Figure 9. Example 2: simulation of a wetting-loading process.

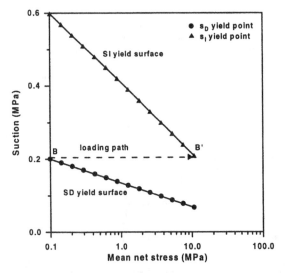

Figure 10. Shape of SD and SI yield surface.

on the right side of MC_I is unattainable, the soil state is pushed by the main drying curve, causing a decrease in water ratio at constant suction. At the end of the loading process, the current state lays at point C, on the main drying curve corresponding to the final value of void ratio. Along the path BB', a decrease in e minus e_w takes place. Since e minus e_w is equal to the volume of air divided by the volume of solid, its decrease indicates an expulsion of air from the sample during the loading step BB'. Afterwards, a decrease in e_w occurs along the path B'C, indicating a subsequent expulsion of water. The sequence: air expulsion followed by water expulsion during loading is typical of a soil compaction process. Example 2 illustrates how the proposed model can represent such a feature.

In Figure 10, the changes in s_I and s_D with the mean net stress along the elastic path BB' are plotted in the $p"$-s diagram. They depict the shape of SI and SD, respectively. Both of them are inclined with respect to the horizontal axis. Each angle expresses the dependence of the corresponding main curve on the void ratio. SI angle is different from SD angle. The inclinations of SI and SD surfaces explain why, as pointed out by Chen et al. (1999), the *apparent maximum ever experienced suction* and the *apparent minimum ever experienced suction* of the soil change with load. SI and SD define a physically based closure of the elastic domain in the $p"$-s diagram.

6 CONCLUSIONS

A coupled elastoplastic hydro-mechanical model has been proposed. It aims at including reversible and irreversible changes in water content and volumetric deformation under stress and suction changes, in a comprehensive framework. The mechanical part of the model is based on the Barcelona Basic Model developed by Alonso et al. (1990). The hydraulic part is formulated on basis of the following:

1. The work-conjugated variable associated to suction s is the opposite of the variation of volumetric water content. The kinematical variable related to suction is the water ratio, defined as the volume of water divided by the volume of solid.
2. The retention curve is split into two parts. The first part is related to the water distribution inside the macropores and it is sensitive to changes in the void ratio. The second part is related to the water storage inside the micropores and it is not affected by the deformation of the soil skeleton.
3. By drawing the macroscopic part of the retention curve for all void ratios in the e-e_w-s diagram, a stable state boundary surface, separating the zone of possible hydraulic states from the zone of impossible hydraulic states, is obtained.
4. Using the concept of state boundary surface, the hysteretic character of changes in water content upon suction cycles can be represented within an elastoplastic framework. This formalism allows moreover for a coupling with mechanical actions, which comes naturally from the dependence of the retention curve on the void ratio.

The model developed gives an insight into the coupled evolution of water ratio and void ratio, such as irreversible changes in water content under mechanical actions, changes in apparent maximum ever experienced suction and apparent minimum ever experienced suction with load, and air-water expulsion sequence during compaction.

ACKNOWLEDGEMENTS

The support of the D.G.E.S.I.C. and of the MURST, through Italy-Spain Integrated Actions Research Grant HI-1998-0119 is gratefully acknowledged. The authors thank A. Di Mariano and J. Alcoverro for their help during the preparation of the paper.

REFERENCES

Alonso, E.E., A. Gens & A. Josa 1990. A constitutive model for partially saturated soils. *Géotechnique*, 40(3):405-430.
Barbour, S.L. 1998. Ninetheenth Canadian Geotechnical Colloquium: The soil-water characteristics curve – A historical perspective. *Can Geotech. J.* 35:873-894.
Chen, Z.-H., D.G. Fredlund & J.K.-M. Gan 1999. Overall volume change, water volume change and yield associated with an unsaturated compacted loess. *Can. Geotech. J.* 36:321-329.
Childs, E. C. 1940. The use of soil moisture characteristics in soil studies. *Journal of Soil Science*, 50: 239-252.
Dangla, P., L. Malinsky & O. Coussy 1997. Plasticity and imbibition-drainage curves for unsaturated soils: A unified approach. *Proc. 6th Int. Symp. on Numerical Models in Geomechanics, Montreal, 2-4 July 1997*: 141-146 Rotterdam: Balkema.
Edgar, T.V. 1993. One and three dimensional, three phase deformation in soil. *Unsaturated Soils Geotechnical Special Publications N° 39*: 139-150 ASCE.
Houlsby, G.T 1997. The work input to an unsaturated granular material. *Géotechnique*, 47(1):193-196.
Nitao, J.J. & J. Bear 1996. Potentials and their role in transport in porous media. *Water Resources Res.* 32(2):225-250.

Rampino, C., C. Mancuso & F. Vinale 1999. Laboratory testing on an unsaturated soil: equipment, procedures, and first experimental results. *Can. Geotech. J.* 36(1):1-12.
Romero, E. 1999. Characterization and thermo-hydro-mechanical behaviour of unsaturated Boom clay:
an experimental study. *Doctoral thesis, Universitat Politècnica de Catalunya, Barcelona, Spain.*
Romero, E., A. Gens, & A. Lloret 1999. Water permeability, water retention and microstructure of un-
saturated Boom clay. *Engineering Geology,* vol. 54, no 1-2, pp. 117-127.
Romero, E. & J. Vaunat 2000. Retention curves in deformable clays. *Proc. International Workshop On
Unsaturated Soils: Experimental Evidence And Theoretical Approaches, Trento: 10-12 April 2000:*
this issue Rotterdam: Balkema.
Vanapalli, S.K., D.G. Fredlund & D.E Pufahl 1999. The influence of soil structure and stress history on
the soil-water characteristics of a compacted till. *Géotechnique,* 49(2):143-159.
Wheeler, S.J. 1996. Inclusion of specific water volume within an elastoplastic model for unsaturated soil.
Can. Geotech. J. 33:42-57.
Wheeler, S.J. & V. Sivakumar 1995. An elastoplastic critical state framework for unsaturated soil.
Géotechnique 45(1):35-53.
Wheeler, S.J. & D. Karube 1996. Constitutive modelling. *Proc. 1ˢᵗ Int. Conf. On Unsaturated Soils,
Balkema, Paris, 6-8 September 1995:* 1323-1356 Rotterdam: Balkema.

Experimental Evidence and Theoretical Approaches in Unsaturated Soils, Tarantino & Mancuso (eds)
© *2000 Taylor & Francis, ISBN 90 5809 186 4*

Remarks on the constitutive modelling of unsaturated soils

C.Jommi
Dipartimento di Ingegneria Strutturale, Politecnico di Milano, Italy

ABSTRACT: Some common ideas concerning the *correct way* to model the constitutive behaviour of unsaturated soils are discussed. Neither theoretical considerations nor experimental evidence may support definite conclusions on the stress state variables that should be adopted in the formulation of constitutive laws. Different approaches may be followed in the constitutive modelling, each of them presenting advantages and shortcomings. Constitutive models for unsaturated soils are usually formulated in terms of two separate stress variable. However, the *average soil skeleton stress*, defined as the difference between total stress and an *equivalent fluid pressure* with saturation degree as weighing parameter, may be equally adopted advantageously. The latter choice allows for a natural reproduction of some aspect of the overall mechanical behaviour. Nevertheless, additional constitutive parameters are usually necessary to model the increase in the preconsolidation stress with suction and collapse upon wetting. The main criteria for the development of a model written in terms of a single stress variable are recalled. These criteria are adopted to adapt Modified Cam Clay for unsaturated states. A comparison is then presented between the latter model and the one proposed by Alonso et al. (1990).

1 INTRODUCTION

In the development of constitutive models for soils in the unsaturated range, the choice of stress state variables has always played a central role. As the principle of effective stress had provided the key for analysing and modelling the mechanical behaviour of saturated soils, in 1959 Bishop proposed to extend its definition to the unsaturated range in the following way:

$$\sigma'_{hk} = \sigma_{hk} - u_a \, \delta_{hk} - \chi \, (u_w - u_a) \, \delta_{hk} \qquad (1)$$

where σ'_{hk} is defined *effective* stress, σ_{hk} is the total stress, u_a is the pressure of the gas and vapour phase, and u_w is the pressure of the liquid phase. Bishop proposed that the weighing parameter χ should be equal to one for saturated soils and zero for dry granular soils. In the intermediate saturation range, a value between one and zero should be expected. The proposed parameter was introduced to take into account the magnitude of the surface tension effects on the overall unsaturated soil behaviour.

Assuming the validity of equation (1), early studies concentrated on the experimental derivation of a value for χ, which could account for the effects of surface tension both on the shear strength and on the volumetric strains. It soon became clear, anyway, that a similar parameter could not be defined for collapsible soils (Jennings & Burland 1962).

The experimental results pushed towards the adoption of two independent stress state variables (Coleman 1962, Bishop & Blight 1963, Blight 1967). Matyas & Radakrishna (1968) developed *state surfaces* for the void ratio e and the saturation degree S_r, for either isotropic or one-dimensional loading of unsaturated soils, adopting two independent stress variables. Fredlund & Morgenstern (1977) presented the results of null tests supporting this choice, and concluded that any two of the three stress variables $(\sigma - u_a)$, $(\sigma - u_w)$, $(u_a - u_w)$ could be sufficient to fully represent the stress state. In the latter expressions, σ may indicate either the volumetric component of the total stress, or the vertical stress in one-dimensional compression. The two stress variables $(\sigma - u_a)$ and $(u_a - u_w)$ are commonly adopted, mainly because in most practical problems the air pressure may be considered constant and equal to the atmospheric pressure. Moreover, even if the adoption of the axis translation technique allows for positive values of both air and water pressures, in the laboratory experimental investigations it is preferable to vary water pressure. Keeping the air pressure constant reduces the influence of the air compressibility on the test data. Results of further null tests, suggesting the dependence of the soil behaviour on two stress variables, both at high and medium saturation degrees, have been recently presented (Tarantino et al. 2000, Tarantino & Mongiovì 2000).

Null tests have been performed in order to demonstrate that two stress variables effectively control the behaviour of unsaturated soils. In principle, similar tests could be thought in order to verify whether a single stress variable could be sufficient. From the experimental point of view, this procedure would be practically impossible and, from a theoretical standpoint, it seems to be useless.

It must be clearly pointed out, in fact, that the choice of the stress variables in the constitutive modelling is a completely different problem from the choice of the controlling variables in experimental investigations, as highlighted for example by Tarantino & Mongiovì (2000). The stress variables used in the modelling of the coupled hydro-mechanical behaviour of unsaturated soils may be totally arbitrary. Any convenient choice may thus be allowed, provided its mechanical consistency.

It is now widely accepted that the constitutive laws for unsaturated soils may be formulated advantageously in the framework of elastoplasticity, and different models have been proposed in the last decade. All of them are constructed extending constitutive laws for saturated soils. Some of them, following the work by Alonso et al. (1990), are written in terms of two separate stress variables, while in other models a single stress variable is substituted for the saturated effective stress. In this second case, however, some modifications to the original saturated laws have been introduced, in order to describe some aspects of the coupled hydro-mechanical unsaturated behaviour.

A comparison between the different models proposed has been presented by Gens (1996) and Wheeler & Karube (1996). In the following, only some considerations on the different possible approaches will be suggested, trying to show the advantages and the shortcomings of the adoption of a single stress variable in the constitutive laws for unsaturated soils.

The main criteria for the development of a model written in terms of a single stress variable are recalled. These criteria are adopted to adapt Modified Cam Clay for unsaturated states. A comparison is then presented between the latter model and the one proposed by Alonso et al. (1990).

2 STRESS VARIABLES IN CONSTITUTIVE MODELLING

The choice of one or two stress variables in the modelling the behaviour of unsaturated soils has been sometimes supported by theoretical considerations on the hypothetical action of suction on the soil skeleton at the micro level, at the contacts between solid particles (Kogho et al. 1993, Gudehus 1995, Wheeler & Karube 1996, Rampino et al. 1999). Although interesting, this approach does not seem to be a way to derive definite conclusions on the topic, as many different mechanisms may control the overall hydro-mechanical response of different soils in the unsaturated state. Moreover, in the experimental tests, generalised stresses and strains may be imposed only at the boundary of the soil specimen and no measures may be performed inside the soil sample. The way in which the mechanisms controlling the unsaturated behaviour are activated by the boundary constraints is not yet fully understood. A constitutive model may thus be postulated just as a *black box* relating generalised static variables to generalised kinematic variables.

The possible ways in which an elastoplastic constitutive model for unsaturated soil may be constructed can be summarised as follows. Consider a generic constitutive law for a saturated soil written in terms of effective stress, in incremental form:

$$\dot{\varepsilon}_{ij} = C^{ep}_{ijhk} \dot{\sigma}'_{hk} \tag{2}$$

where $\dot{\varepsilon}_{ij}$ is the strain rate, $\dot{\sigma}'_{hk}$ is the effective stress rate and C^{ep}_{ijhk} is the tangent elastoplastic matrix in saturated conditions. Models written in terms of two stress variables usually associate the saturated elastoplastic matrix to net stress increments, and add an explicit dependence on suction:

$$\dot{\varepsilon}_{ij} = C^{ep}_{ijhk}(\dot{\sigma}_{hk} - \dot{u}_a \delta_{hk}) + C^{s}_{ijhk}(\dot{u}_a - \dot{u}_w)\delta_{hk} \tag{3}$$

where C^{s}_{ijhk} is a new constitutive matrix, relating strain increments to suction increments $(\dot{u}_a - \dot{u}_w)\delta_{hk}$. Adopting this approach, the effects of net stress and suction on the overall mechanical behaviour are separated, and a complete new set of parameters must be introduced to define the dependence of the constitutive behaviour on the suction.

Different single stress variables have been proposed, but, in this work, the attention will be focused only on a particular form, denoted by $\hat{\sigma}_{hk}$, which can be called *average soil skeleton stress*. This stress variable is defined as the difference between the total stress and the mean value of the fluid pressures weighted with the saturation degree S_r:

$$\hat{\sigma}_{hk} = \sigma_{hk} - [S_r u_w + (1 - S_r)u_a]\delta_{hk} . \tag{4}$$

It is worthwhile noting that, contrary to a common statement, the saturation degree should not be considered a constitutive parameter, as it represents a volume fraction. The average soil skeleton stress may be rewritten in the following way:

$$\hat{\sigma}_{hk} = (\sigma_{hk} - u_a \delta_{hk}) + S_r(u_a - u_w)\delta_{hk} \tag{5}$$

highlighting its dependence on net stress and suction.

By simply substituting the average soil skeleton stress for effective stress in a constitutive model written for saturated soils (Eq. 2), the following relation is obtained:

$$\dot{\varepsilon}_{ij} = C^{ep}_{ijhk} \dot{\hat{\sigma}}_{hk} , \tag{6}$$

which, rewritten in terms of net stress and suction, gives:

$$\dot{\varepsilon}_{ij} = C^{ep}_{ijhk}(\dot{\sigma}_{hk} - \dot{u}_a \delta_{hk}) + C^{ep}_{ijhk}[S_r(\dot{u}_a - \dot{u}_w)\delta_{hk}] + C^{ep}_{ijhk}[\dot{S}_r(u_a - u_w)\delta_{hk}]. \tag{7}$$

A dependence of the overall behaviour on suction and saturation degree is included in the preceding equation, without the introduction of any new constitutive parameters.

The saturation degree increment may be expressed as a function of suction variation as:

$$\dot{S}_r = \frac{\partial S_r}{\partial(u_a - u_w)}(\dot{u}_a - \dot{u}_w) \tag{8}$$

which, substituted in equation (7), gives:

$$\dot{\varepsilon}_{ij} = C^{ep}_{ijhk}(\dot{\sigma}_{hk} - \dot{u}_a \delta_{hk}) + C^{ep}_{ijhk}\left[S_r + \frac{\partial S_r}{\partial(u_a - u_w)}(u_a - u_w)\right](\dot{u}_a - \dot{u}_w)\delta_{hk} . \tag{9}$$

The comparison between equation (3) and equation (9) shows that the compliance tensor, relating strain increments to suction increments, is given by the expression:

$$C^{s}_{ijhk} = C^{ep}_{ijhk}\left[S_r + \frac{\partial S_r}{\partial(u_a - u_w)}(u_a - u_w)\right] . \tag{10}$$

141

A detailed comparison between the two formulations will be presented in Section 5, with reference to the Modified Cam Clay as basic saturated elastoplastic model.

Here, it is worthwhile noting that in the definition of the average soil skeleton stress (Eq.5) the contribution of suction may be considered equivalent to an increment in the isotropic stress. It is obvious that the simple introduction of this single stress variable in a constitutive law for saturated soils will reproduce, at the macroscopic level, the effects which may be associated to variations in the volumetric stress. The experimental evidence shows that, starting from saturated conditions, an increase in suction produces initially an increase in the shear strength and an increase in both the shear and the volumetric stiffness. These aspects of the behaviour may be effectively associated to an increment in the isotropic stress, and they may be qualitatively reproduced just be adopting the average soil skeleton stress.

Nevertheless, for collapsible soils, an increase in suction produces an increase in the apparent preconsolidation stress in the net stress-void ratio plane. This aspect of the unsaturated behaviour cannot be reproduced just by substituting the average soil skeleton stress for effective stress in a constitutive model written for saturated conditions. The prediction of the model would be exactly the opposite, which means a decrease in the apparent preconsolidation stress with suction.

Upon wetting the soil may experience swelling and/or collapse as a function of the net stress level. It is frequently claimed that a single stress variable should not be adopted in the modelling of the unsaturated behaviour, as it does not allow reproducing the collapsing behaviour. The collapsing behaviour of an unsaturated soil, in reality, is the macroscopic evidence of a structural instability of the soil skeleton, and it is totally independent of the stress variables adopted in the constitutive modelling.

In order to clarify this point, the relevant experimental data of triaxial tests performed by Matiotti et al. (1995) may be recalled. Saturated loose sand samples where tested in triaxial conditions, following a conventional compression path up to different stress ratios. The effective isotropic stress was subsequently reduced, at constant deviatoric stress. As the isotropic component of stress is reduced, an increase in the void ratio would be expected. The experimental data show that the volumetric behaviour is dependent on the stress level. For low stress levels, an increase in void ratio is observed, while for higher stress levels the soil samples undergo compaction with an increase in relative density. Matiotti et al. interpreted this collapsible behaviour as a structural instability, and modelled the experimental data in a conventional elastoplastic framework adopting the usual effective stress tensor.

In principle, no reasons support the idea that a single stress variable should not be used in constitutive modelling. However, the increase in the apparent preconsolidation stress with suction, and collapse upon wetting, cannot be interpreted as an effect of a mere increase and decrease of the mean pressure acting on the soil skeleton. These effects are more similar, from a macroscopic point of view, to the effects of bonding and debonding due to a *cementing* action. A constitutive model written for saturated conditions must be necessarily modified in order to reproduce these characteristics of the overall behaviour.

In fact, no single stress variable has ever been found which, substituted for effective stress, allows for a description of *all* the aspects of the mechanical behaviour of a given soil in the unsaturated range, without modifications in the constitutive model of the same saturated soil. In other words, the adoption of a single stress variable of the type defined in equation (5) allows for a natural description of the aspects of the overall behaviour linked to a variation in the mean stress acting on the soil skeleton. Besides, the *cementing* effect due to suction can be introduced in the constitutive model, only by modifying the basic saturated one. This path of reasoning is supported by some recent works, which will be recalled in the next section.

3 STRESS AND CONJUGATE STRAIN VARIABLES

Although for long time the debate has been concentrated on the choice of stress variables, in the definition of a *correct* framework, for the development of constitutive models, strain variables should be taken into account too. Wheeler & Sivakumar (1995) have apparently introduced the latter concept, and Houlsby (1997) presented the first simple theoretical analysis on the topic.

Hassanizadeh & Gray (1990), Hutter et al. (1999), Muraleetharan & Wei (1999), among others, proposed full thermodynamic analyses of multiphase deformation.

Houlsby pointed out that the choice of stress and strain variables is arbitrary, provided that the variables adopted are *work conjugate*. This guarantees that the rate of input work per unit volume of the soil equates the sum of the products of the generalised stresses with their corresponding strain rates. With reference to undrained conditions, neglecting the contribution of the work dissipated by the flow of water and air through the soil, the expression proposed by Houlsby (1997) for the rate of input work $\dot{W}$ (per unit volume of soil skeleton) may be written as:

$$\dot{W} = u_a\, n(1-S_r)\,\dot{\rho}_a/\rho_a - (u_a - u_w)\,n\dot{S}_r + \left\{ \sigma_{hk} - [S_r\, u_w + (1-S_r)\,u_a]\delta_{hk} \right\}\dot{\varepsilon}_{hk}. \quad (11)$$

In equation (11), the first term is the rate of input work to compress the air phase, with ρ_a being the air density and n the soil skeleton porosity. The second term represents the rate of input work necessary to change the saturation degree, and shows that the variation of the volume of water per unit volume of soil skeleton is the generalised strain conjugate to suction. The last term shows that the static variable conjugate to the soil skeleton strain is the average soil skeleton stress (Eq.5).

By rearranging the second and the third term in equation (11), other combinations of conjugate stresses and strains may be equally defined. For example, if the net stress is associated to the soil skeleton strain, the generalised strain quantity $(-n\dot{S}_r + S_r\varepsilon_{kk})$ must be associated to suction (see e.g. Romero & Vaunat 2000).

In the derivation of equation (11), Houlsby neglected the work dissipated by the *air-water interface*, assuming that the relative velocity between the soil skeleton and the interface itself vanishes. This hypothesis seems to be equivalent to the assumption that the saturation degree remains constant, in contrast with the presence of the second term in equation (11). Although Houlsby's proposal seems to present this inconsistency, another point of view may be chosen, in order to clarify this point. Dangla et al. (1997) presented a similar analysis, coming to the same conclusion with respect to conjugate stress and strain variable. However they stated that, in the case of unsaturated soils, the *soil skeleton* must be defined as *the system composed of both solid grains and the phase interfaces*.

This interpretation suggests that, when the average soil skeleton stress is adopted as a single stress variable in the development of a constitutive law, the latter must describe the behaviour of the solid grains packing together with the action of the interfaces. It becomes clear, therefore, that a possible consistent way to construct a model for unsaturated soils, starting from a saturated one, may be summarised in the two following steps:
– substitution of the average soil skeleton stress (Eq. 5) for effective stress.
– introduction in the basic saturated elastoplastic model of the modifications necessary to take into account the effects of the interfaces on the overall mechanical behaviour.

The substitution of the average soil skeleton stress for effective stress takes into account the effect of suction on the mean volumetric stress acting on the soil skeleton. Thus, it seems logical to associate the action of the interfaces on the overall mechanical behaviour to the macroscopic bonding and debonding effect.

It is reasonable to think that the two effects of suction may prevail in turn, for different soils in different suction ranges. If its influence on the mean volumetric stress is prevalent, the original basic saturated model may be sufficient to reproduce the macroscopic behaviour. In this case, the adoption of a single stress variable may be advantageous. On the contrary, if the action of the interfaces mainly governs the overall behaviour, the basic saturated model must be modified, with the introduction of new parameters. If the number of necessary parameters increases significantly, the use of a single stress variable, instead of net stress and suction, becomes less advantageous.

One of the shortcomings of the proposed method is that, although conceptually simple, it may be difficult to give explicit analytical expression of the constitutive laws in terms of net stress and suction. The difficulty arises from the product between saturation degree and suction appearing in the definition of the average soil skeleton stress and from the non-linear relationship between the two variables.

The main disadvantage of the procedure is that the saturation degree cannot be measured in experimental tests. It must be estimated from the measures of both water content variation and volume variation. This fact reduces substantially the possibility of comparing published experimental data with the model prediction. In fact, besides the data of the mechanical tests, the soil water characteristic curve (SWCC) must be known, in order to fully describe the coupled hydromechanical behaviour.

4 MODIFIED CAM CLAY FOR UNSATURATED STATES

The procedure outlined before may be applied to any elastoplastic constitutive model (see e.g. Jommi & di Prisco 1994, Bolzon et al. 1996). It will be followed here starting from the Modified Cam Clay model (Roscoe & Burland, 1968), in order to compare the formulation with the well-known Basic Barcelona Model (BBM) proposed by Alonso et al. (1990).

In the following the usual triaxial stress and strain variables will be adopted:
- $p = (\acute{\sigma}_1 + 2\acute{\sigma}_3)/3 = $ total mean stress
- $q = (\acute{\sigma}_1 - \acute{\sigma}_3) = $ deviatoric stress
- $\mathring{\varepsilon}_v = (\mathring{\varepsilon}_1 + 2\mathring{\varepsilon}_3) = $ volumetric strain, sum of the elastic and plastic parts: $\mathring{\varepsilon}_v = \mathring{\varepsilon}_v^e + \mathring{\varepsilon}_v^p$
- $\mathring{\varepsilon}_s = 2(\mathring{\varepsilon}_1 - \mathring{\varepsilon}_3)/3 = $ deviatoric strain, sum of the elastic and plastic parts: $\mathring{\varepsilon}_s = \mathring{\varepsilon}_s^e + \mathring{\varepsilon}_s^p$
- $s = (u_a - u_w) = $ matric suction
- $p' = p - u_w = $ effective mean stress
- $p'' = p - u_a = $ net mean stress
- $\hat{p} = p'' + S_r s = $ average soil skeleton mean stress
- $v = $ specific volume $= 1 + e$, with e denoting the void ratio.

4.1 Isotropic stress states

Consider first the behaviour under isotropic compression, represented in Figure 1, in a semilogarithmic plot relating net mean stress to specific volume. The virgin compression line for saturated conditions, with $s = 0$, is represented by the line A-A', and it is given by the following expression:

$$v = N - \lambda \ln \frac{p'}{p_c} = N(0) - \lambda(0) \ln \frac{p''}{p_c}. \tag{12}$$

In equation (12) $N = N(0)$ and $\lambda = \lambda(0)$ are the usual Cam Clay parameters. Note that in saturated conditions p'' coincides with the effective stress p'. By simple substituting the average soil skeleton stress for p' in equation (12), the unsaturated compression curve is obtained:

$$v = N(s) - \lambda(0) \ln \frac{\hat{p}}{\hat{p}_c}. \tag{13}$$

Projected in the (v, p') plane, for constant values of suction, the compression curve assumes the shape indicated by B-B' in Figure 1.

Consider, in fact, an increase in suction at constant net stress, starting from saturated conditions at point A. The increase in suction will induce an increase in the average soil skeleton mean stress and, consequently, a reduction in the specific volume. As the point A lays on the virgin compression line, an increase in $\hat{p}$ will induce an elastoplastic reduction in the specific volume

$$v = N(s) = N(0) - \lambda(0) \ln \frac{\hat{p}_B}{\hat{p}_A} = N(0) - \lambda(0) \ln \frac{p_c + S_r s}{p_c}, \tag{14}$$

and the normal compression curve will then start from point B.

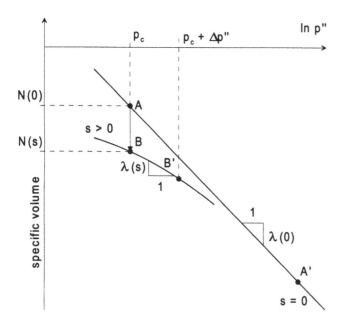

Figure 1. Isotropic compression curves projected in the net stress - specific volume plane

If an increase in net stress at constant suction is now considered, the specific volume will change, following the incremental law:

$$\dot{v} = -\lambda(0) \frac{\dot{\hat{p}}}{\hat{p}}. \tag{15}$$

As $\dot{\hat{p}} = \dot{p}''$ at constant suction, equation (15) may be rewritten in the following way:

$$\dot{v} = -\lambda(0) \frac{\dot{p}''}{\hat{p}} = -\lambda(0) \frac{\dot{p}''}{(p'' + S_r s)} = -\left[\frac{\lambda(0) p''}{(p'' + S_r s)} \right] \frac{\dot{p}''}{p''}. \tag{16}$$

Equation (16) shows that the isotropic stiffness increases with suction. The slope of the compression curve in the (v, p'') plane at constant suction may be denoted as $\lambda(s)$, and from equation (16) it is immediate to derive that:

$$\lambda(s) = \left[\frac{\lambda(0) p''}{(p'' + S_r s)} \right]. \tag{17}$$

It is also evident from equation (17) that the slope of the compression curve, for any value of suction, is smaller than the slope of the saturated compression line. It increases with increasing net stress, and the compression curve is thus not linear in this plane (B-B' in Fig.1). The linearity of the compression curve is maintained in the $(v, \hat{p})$ plane, as equation (13) clearly shows.

Similar considerations hold for the elastic part of the volumetric strain. In the original saturated model it is given by:

$$\dot{\varepsilon}_v^e = \frac{\kappa}{v} \frac{\dot{p}'}{p'} = \frac{\kappa(0)}{v} \frac{\dot{p}''}{p''}. \tag{18}$$

where $\kappa = \kappa\,(0)$ is the Cam Clay elastic volumetric compliance. By substitution of the average soil skeleton stress for effective stress, the following incremental law for the elastic volumetric strain is obtained:

$$\dot{\varepsilon}_v^e = \frac{\kappa(0)}{v}\,\frac{\dot{\hat{p}}}{\hat{p}}. \tag{19}$$

Again, considering constant suction paths, the preceding expression may be written as:

$$\dot{\varepsilon}_v^e = \kappa(0)\,\frac{\dot{p}''}{\hat{p}} = \kappa(0)\,\frac{\dot{p}''}{(p''+S_r s)} = \kappa(s)\frac{\dot{p}''}{p''} = \left[\frac{\kappa(0)p''}{(p''+S_r s)}\right]\frac{\dot{p}''}{p''}. \tag{20}$$

Equation (20) shows that the model predicts an increase in the elastic stiffness with suction, and its non linear decrease with net stress, at constant suction. The unloading-reloading path is linear in the $(v,\hat{p})$ plane.

All the relevant characteristics of the model highlighted in this paragraph are qualitatively consistent with the experimentally observed behaviour. No new parameters have been added to the ones adopted for saturated conditions in the constitutive model for the soil skeleton.

Nevertheless, it has to be pointed out that the model, at this stage, predicts a decrease in the apparent preconsolidation stress with suction in the (v, p'') plane, as will be shown more clearly in the next paragraph. Some swelling soils may present this kind of behaviour, as reported for example by Aitchison & Woodburn (1969). On the contrary, in many cases an increase in the apparent preconsolidation stress is observed. Only a modification in the saturated model allows reproducing this characteristic of the behaviour, as will be shown in paragraph 4.3.

4.2 *Triaxial stress states*

The yield function for saturated states is written as:

$$f = q^2 - M^2 p'(p_0^* - p') = 0 \tag{21}$$

where M is the slope of the conventional critical state line (related to the friction angle), and p_0^* is the isotropic hardening parameter, representing the preconsolidation stress. Substituting the average soil skeleton stress for the effective stress in equation (21), the yield function for unsaturated conditions is obtained

$$f = q^2 - M^2 \hat{p}(p_0^* - \hat{p}) = 0, \tag{22}$$

which may be rewritten in terms of net stress and suction, giving

$$f = q^2 - M^2(p''+S_r s)(p_0^* - p'' - S_r s) = 0. \tag{23}$$

Equation (23) shows that the projection of the yield surface in the (p'', s) plane, for a constant value of suction, is an ellipse of the same shape and the same dimensions as the saturated one, translated backwards by a quantity equal to $S_r s$. The yield function is represented in Figure 2 in the (p'', q) plane, together with the corresponding saturated one. In the figure, the projections of the critical state lines are depicted too. Writing the expression of the critical state line in terms of the average soil skeleton stress, the following equation is obtained

$$f = M\,\hat{p} = M\,p''+M\,S_r s, \tag{24}$$

which shows that the model predicts an increase in the apparent cohesion with suction. The slope of the critical state line, and thus the friction angle, does not vary.

Rearranging the terms in equation (23), the following form for the yield function is obtained

$$f = q^2 - M^2(p''+S_r s)\left[(p_0^* - S_r s) - p''\right] = q^2 - M^2(p''+S_r s)(\hat{p}_0 - p'') = 0 \tag{25}$$

which highlights the dependence of the preconsolidation stress on suction.

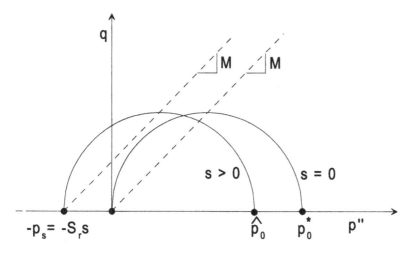

Figure 2. Projection of the yield surfaces in the (p'', q) plane

The intersection with the net stress axis of the projection of the ellipse, for a given value of suction, is denoted by $\hat{p}_0$ in Figures 2,3. Its evolution with suction is depicted in Figure 3, in the (p'', s) plane. From the figure, it is evident that the apparent preconsolidation stress, for any given value of suction, is lower than the corresponding saturated value p_0^*. As already pointed out, this prediction is in contrast with the experimental evidence for collapsing soils. The modification necessary to reproduce an increase in the apparent preconsolidation stress will be presented in the next paragraph.

The elastic part of the deviatoric strain increment may be written, in the saturated model, as

$$\dot{\varepsilon}_s^e = \frac{\dot{q}}{3G}. \tag{26}$$

in which G is the elastic shear modulus. As the deviatoric strain increment is a function of the deviatoric stress only, it is independent of the suction. A dependence on the suction could be introduced only if in the original saturated model the deviatoric elastic strain were a function of the mean stress. In the latter case, an increase in the shear stiffness with suction would be predicted (see e.g. Jommi & di Prisco 1994).

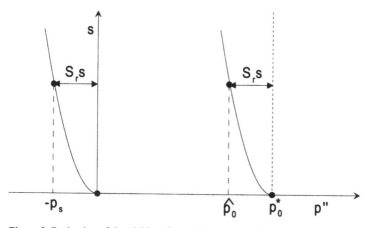

Figure 3. Projection of the yield surfaces in the (p'', s) plane

Alonso et al. (1990) suggest the adoption of a non associated flow rule to avoid overestimating K_0 value. If the proposed flow rule is adopted, written in terms of the average soil skeleton stress, the following equation for the plastic potential is obtained:

$$g = \alpha q^2 - M^2 \hat{p}(p_0^* - \hat{p}) = 0.$$ (27)

The value of α may be related to the slope of the critical state line, imposing zero lateral strains for K_0 conditions. The flow rule does not introduce any additional constitutive parameter.

4.3 Modifications to the saturated model

To reproduce the increase in the apparent preconsolidation stress with suction, and the collapsing behaviour upon wetting, it is sufficient to modify the hardening rule for $\hat{p}_0$. The basic model would give (Eq. 25):

$$\hat{p}_0 = (p_0^* - S_r s).$$ (28)

An additional term, explicitly depending on the saturation degree, may be proposed for the evolution law of the hardening parameter:

$$\hat{p}_0 = (p_0^* - S_r s) + h(S_r)$$ (29)

in which $h(S_r)$ denotes a generic (positive) reversible function of the saturation degree. The function $h(S_r)$ should account for the effect of the interfaces on the overall mechanical behaviour. The proposed modification is conceptually equivalent to the way in which bonding may be introduced in constitutive laws for soils (Gens & Nova, 1993).

In the expression of the additional term, h, the saturation degree should be preferred to the suction, due to the hysteresis in the SWCC. If the hysteresis is neglected, the dependence on the saturation degree implies a unique dependence on suction. Nevertheless, if the hysteresis plays a significant role, it is reasonable to assume that the constitutive behaviour of the soil skeleton depends on the position of the interfaces, thus on the saturation degree, rather than on the suction. The modified yield function is represented in Figures 4,5 in the (p'', q) and in the (p'', s) planes, respectively.

As $h(S_r)$ is a reversible function of suction, it will increase the preconsolidation stress in a drying path, and it will reduce it upon wetting. Consider, in fact, a stress state, represented by point A in Figure 6, initially inside the yield locus (a) for the given value of suction s_a. Consider a wetting path at constant net mean stress. The soil will initially undergo elastic swelling, due to the reduction in the average mean stress. Besides, as saturation increases, the apparent precon-

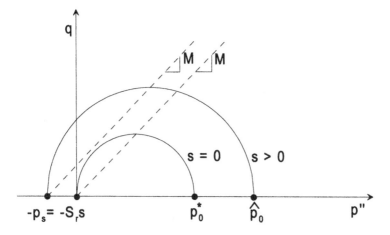

Figure 4. Projection of the modified yield surface in the (p'', q) plane

solidation pressure is reduced, and the yield surface shrinks until it reaches point A. Upon further wetting, the apparent preconsolidation stress does not change, due to the consistency condition that imposes to the stress point A to remain on the yield surface (b). The soil will undergo plastic volumetric compression, and thus *collapse*.

In a drying path, at constant net stress, the soil will experience elastic compression, as the preconsolidation stress shifts onward due to the decrease in the saturation degree. A reloading path will be followed for an increment in the net stress at constant suction, until the new value of preconsolidation stress is reached. Further loading will induce elastoplastic compression.

The evolution law h (S_r) must be derived on the bases of the experimental data, and its definition requires the introduction of new constitutive parameters.

It is worthwhile noting that, starting from saturated conditions, the evolution law h (S_r) should become active on the onset of desaturation. This means that before the air-entry value is reached, the preconsolidation stress should evolve following the basic law expressed by equation (28). The value of the preconsolidation stress should then follow the dotted line in Figure 5, until the air-entry value is reached. A more detailed discussion on this topic may be found in the work by Geiser et al. (2000).

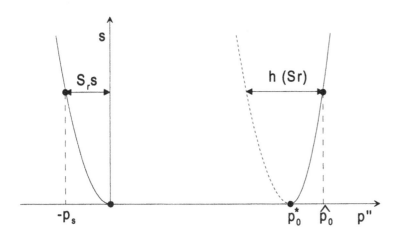

Figure 5. Projection of the modified yield surface in the (p'', s) plane

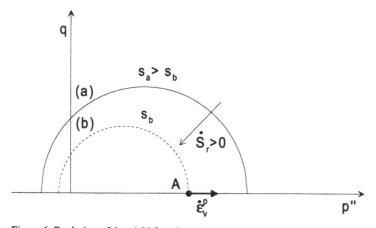

Figure 6. Evolution of the yield function upon wetting at constant net stress.

In the original model, the intersection with the p'' axis of the yield surface on the traction side, denoted by p_s, is equal to $S_r s$. In principle, p_s may evolve with the saturation degree in a different way, as $\hat{p}_0$ does. In this case, a new evolution function for p_s should be provided. The simple basic translation implies an increase in the apparent cohesion proportional to $S_r s$, and a constant friction angle, as equation (24) shows. Many published experimental data (see e.g. Vanapalli et al. 1996) have been modelled by similar expressions, thus suggesting that no modifications may be necessary for the evolution of p_s.

5 COMPARISON BETWEEN THE PROPOSED MODEL AND THE BBM

In the Barcelona Basic Model (BBM), the constitutive parameters which define the saturated elastoplastic behaviour (G, λ, κ, N, M) are associated to the net stress (with the notation introduced in paragraph 4.1). A set of new parameters (r, β, κ_s, p^c, k) is defined in order to describe the influence of the suction on the overall mechanical behaviour. The parameters r and β describe the variation of the slope of the normal compression line, $\lambda(s)$, with suction, and κ_s is the elastic stiffness for changes in suction. The reference pressure, p^c, somehow relates the specific volume for a given value of suction, $N(s)$, to the specific volume in the saturated conditions $N = N(0)$. The last parameter, k, describes the evolution of the intersection of the yield function with the p'' axis, on the traction side, and governs the increase in the apparent cohesion.

In order to compare the two formulations, the relevant expressions of the BBM and of the model proposed, adopting a single stress variable (SSM) will be summarised in the following.

The slope of the compression curves in the SSM (Eq. 17) is given by

$$\lambda(s) = \lambda(0)\left[p''(p''+S_r s)^{-1}\right], \tag{30a}$$

while in the BBM it is a function of suction only:

$$\lambda(s) = \lambda(0)\left[(1-r)\exp(-\beta s)+r\right]. \tag{30b}$$

The comparison between equations (30a,b) clearly shows the advantages and the shortcomings of the two approaches. The behaviour predicted by the SSM is qualitatively similar to the one proposed by the BBM. While in the latter the effects of suction and net stress may be completely separated, in the SSM a coupled dependence of $\lambda(s)$ on suction and net stress is predicted. No additional constitutive parameters are required for the mechanical part of the model. Nevertheless, it is important to note that the relationship between the saturation degree and the suction (SWCC) must be known. The hydraulic parameters necessary to define the SWCC are implicitly added to the model. This aspect represents its major disadvantage from the practical point of view, although, from a theoretical standpoint, the SWCC is essential to complete the description of the coupled hydro-mechanical behaviour (Buisson & Wheeler 2000, Vaunat et al. 2000). The absence of *ad hoc* constitutive parameters gives less flexibility to the model. Nevertheless, it is obvious that the scope of the presented approach is exactly to reduce the number of constitutive parameters.

The proposed observations on $\lambda(s)$ hold true for the other relevant volumetric quantities. The elastic compliance in the SSM has an expression analogous to the elastoplastic one:

$$\kappa(s) = \kappa(0)\left[p''(p''+S_r s)^{-1}\right]. \tag{31a}$$

In the BBM κ is assumed to be constant and equal to the saturated value

$$\kappa(s) = \kappa(0) = \kappa. \tag{31b}$$

The yield function (Eq.25) may be written as:

$$f = q^2 - M^2(p''+S_r s)(\hat{p}_0 - p'') = 0. \tag{32a}$$

The evolution law for the preconsolidation stress is given in equation (29).

In the BBM, the yield function takes the form:

$$f = q^2 - M^2(p''+ks)(p_0 - p'') = 0 \tag{32b}$$

150

with the preconsolidation stress given by:

$$\left(\frac{p_0}{p^c}\right)=\left(\frac{p_0^*}{p^c}\right)^{\frac{(\lambda(0)-\kappa)}{(\lambda(s)-\kappa)}}.$$ (33)

The comparison between the equation (29) and equation (33) shows that the evolution laws for the preconsolidation stress are formally similar. While in the BBM it is a function of the unsaturated parameters r, β, e p^c, in the SSM it depends on the parameters necessary to describe the function h (S_r).

It is worthwhile noting that the evolution law proposed in BBM might be forced in the SSM model, adopting the following expression:

$$\hat{p}_0 = p^c\left(\frac{p_0^*}{p^c}\right)^{\frac{(\lambda(0)-\kappa)}{(\lambda(s)-\kappa)}}+S_r s.$$ (34)

The comparison between Equations (29) and (34) shows that, in this case, the additional function h should depend on both the saturation degree and the suction, having the formal expression:

$$h = p^c\left(\frac{p_0^*}{p^c}\right)^{\frac{(\lambda(0)-\kappa)}{(\lambda(s)-\kappa)}}+2S_r s + p_0^*.$$ (35)

To force exactly in the SSM the same evolution law for the preconsolidation stress as in BBM, the same parameters r, β, e p^c should be adopted. However, the major advantage of the proposed approach would be lost.

The proposed framework provides a valuable alternative if the overall soil hydro-mechanical behaviour can be modelled with the introduction of a number of new constitutive parameters significantly smaller than that characterising a model of the BBM type. However, if the quantitative predictions of the model are far from the experimental evidence, more constitutive parameters must be added to improve its performance. In this case, this approach becomes equivalent to the traditional one, and thus useless.

A possible advantage, which may be exploited in the future when more experimental data will be available, is that the effects of the hysteresis in the SWCC may be already built in the model naturally.

It is usually claimed that one of the advantages in the adoption of a single stress variable is that it *provides a natural transition* between the unsaturated and the saturated states. In the author's opinion, this is a misleading concept. On the one hand, in fact, the mathematical transition between the two states may be equally accommodated in the two cases (Vaunat et al. 1997). On the other hand, the transition between unsaturated and saturated states is a physical process much more complicated than a simple *switching* of the stress variables.

6 CONCLUSIONS

Suction may have two effects on the overall mechanical behaviour of an unsaturated soil, at the macroscopic level. On the one hand, it increases the average volumetric stress acting on the soil skeleton. On the other hand, it has a sort of *cementing* effect on the soil packing. The *bonding* action due to a suction increase, and the *debonding* upon wetting, may be the macroscopic evidence of the action of the air-water interface on the soil skeleton.

A valuable single stress variable may be defined as the difference between the total stress and the mean value of the fluid pressures, weighted with the saturation degree. This single stress variable is defined *average soil skeleton stress*. It corresponds to the expression proposed by Bishop with the parameter χ equal to the saturation degree.

By simply substituting the average soil skeleton stress for the effective stress in a constitutive law describing the saturated mechanical behaviour, the first of the two effects is naturally reproduced, without the introduction of any new constitutive parameter. The saturated constitutive law must be modified in order to reproduce the second effect. An explicit dependence of the hardening law for the preconsolidation stress on suction and/or saturation degree is suggested.

The proposed framework has been compared with the traditional approach, in which the constitutive laws are written in terms of net stress and suction separately. It may provide a valuable alternative if a comparable quantitative agreement between the theoretical predictions and the experimental data is obtained with a smaller number of new constitutive parameters.

Nevertheless, the knowledge of the soil water characteristic curve is required, as the saturation degree appears in the definition of the adopted stress variable. This appears to be the main practical shortcoming of the proposed formulation. However, as the SWCC is now receiving increasing attention, this limitation will probably become less severe in the future.

ACKNOWLEDGEMENTS

The support of the MURST, through Italy-Spain Integrated Actions Research Grant is gratefully acknowledged.

REFERENCES

Aitchison, G.D. & J.A. Woodburn 1969. Soil suction in foundation design. *Proc.VII Int. Conf. Soil Mech. Found. Engng.*, Mexico: 1-8.

Alonso, E.E., A. Gens & A. Josa 1990. A constitutive model for partially saturated soils. *Géotechnique*, 40(3): 405-430.

Bishop, A.W. 1959. The principle of effective stress. *Teknisk Ukeblad* 106(39): 859-863.

Bishop, A.W. & G.E. Blight 1963. Some aspects of the effective stress in saturated and partially saturated soils. *Géotechnique*, 13(3): 177-197.

Blight, G.E. 1967. Effective stress evaluation for unsaturated soils. *J. Soil Mech. Found. Div. Am. Soc. Civ. Engrs.* 93(SM2): 125-148.

Bolzon, G., B.A. Schrefler & O.C. Zienkiewicz 1996. Elastoplastic soil constitutive laws generalized to partially saturated states. *Géotechnique* 46(2): 279-289.

Buisson, M. & S. Wheeler 2000. Inclusion of hydraulic hysteresis in a new elasto-plastic framework for unsaturated soils. In *Experimental Evidence and Theoretical Approaches in Unsaturated Soils; Proc. of an International Workshop, Trento, 10-12 April 2000, this issue.*

Coleman, J.D. 1962. Stress-strain relations for partly saturated soils. *Correspondence to Géotechnique* 12(4): 348-350.

Dangla, P., L. Malinsky & O. Coussy 1997. Plasticity and imbibition-drainage curves for unsaturated soils: A unified approach. *Proc. 6th Int. Symp. on Numerical Models in Geomechanics, Montreal, 2-4 July*: 141-146 Rotterdam: Balkema.

Fredlund, D.G. & N.R. Morgenstern 1977. Stress state variables for unsaturated soils. *J. Geotech. Engng. Div. Am. Soc. Civ. Engrs.* 103(GT5): 447-466.

Geiser, F., L. Laloui & L. Vulliet 2000. Modelling the behaviour of unsaturated silt. In *Experimental Evidence and Theoretical Approaches in Unsaturated Soils; Proc. of an International Workshop, Trento, 10-12 April 2000, this issue.*

Gens, A. 1996. Constitutive modelling: Application to compacted soils. *Proc. 1st Int. Conf. On Unsaturated Soils, Paris, 6-8 September 1995*: 1179-1200 Rotterdam: Balkema.

Gens, A. & R. Nova 1993. Conceptual bases for a constitutive model for bonded soils and weak rocks. In *Geotechnical Engineering of Hard Soils – Soft Rocks, Anagnostopoulos et al. (eds.)*: 485-494 Rotterdam: Balkema.

Gudehus, G. 1995. A comprehensive concept for non-saturated granular bodies. *Proc. 1st Int. Conf. On Unsaturated Soils, Paris, 6-8 September 1995*: 725-737 Rotterdam: Balkema.

Hassanizadeh, S.M. & W.G. Gray 1990. Mechanics and thermodynamics of multiphase flow in porous media including interphase boundaries. *Adv. Water Resources* 13:169-186.

Houlsby, G.T. 1997. The work input to an unsaturated granular material. *Géotechnique*, 47(1): 193-196.

Hutter, K., L. Laloui & L. Vulliet 1999. Thermodynamically based mixtures models of saturated and unsaturated soils. *Mech. Cohes.-Frict. Mat.* 4:295-338.

Jennings, J.E.B. & J. B. Burland 1962. Limitations to the use of effective stresses in partly saturated soils. *Géotechnique*, 12(2):125-144.

Jommi, C. & C. di Prisco 1994. A simple theoretical approach for modelling the mechanical behaviour of unsaturated granular soils (in Italian). In *Il ruolo dei fluidi in ingegneria geotecnica; Proc. Italian Conf. Mondovì* 1(II): 167-188.

Kogho, Y., M. Nakano & T. Miyazaki 1993. Theoretical aspects of constitutive modelling for unsaturated soils. *Soils and Foundations* 33(4): 49-63.

Matiotti, R., C. di Prisco & R. Nova 1995. Experimental observations on static liquefaction of loose sands. In *Earthquake Geotechnical Engineering, Ishihara (ed.)*: 817-822 Rotterdam: Balkema

Matyas, E.L. & Radhakrishna 1968. Volume change characteristics of partially saturated solis. *Géotechnique*, 18(4): 432-448.

Muraleetharan, K.K. & C. Wei 1999. Dynamic behaviour of unsaturated porous media: governing equations using the theory of mixtures with interfaces (TMI). *Int. J. Numer. Anal. Meth. Geomech.* 23: 1579-1608.

Rampino, C., C. Mancuso & F. Vinale 1999. Laboratory testing on an unsaturated soil: equipment, procedures, and first experimental results. *Can. Geotech. J.* 36(1): 1-12.

Romero, E. & J. Vaunat 2000. Retention curves in deformable clays. In *Experimental Evidence and Theoretical Approaches in Unsaturated Soils; Proc. of an International Workshop, Trento, 10-12 April 2000, this issue.*

Roscoe, K.H. & J.B. Burland 1968. On the generalised stress-strain behaviour of 'wet' clay. In *Engineering Plasticity*: 535-609 Cambridge: Cambridge University Press.

Tarantino A., & L. Mongiovì 2000. Experimental investigations on the stress variable governing unsaturated soil behaviour at medium to high degrees of saturation. In *Experimental Evidence and Theoretical Approaches in Unsaturated Soils; Proc. of an International Workshop, Trento, 10-12 April 2000, this issue.*

Tarantino, A., L. Mongiovì & G. Bosco 2000. An experimental investigation on the isotropic stress variables for unsaturated soils. *Géotechnique* 50(3): 275-282.

Vanapalli, S.K., D.G. Fredlund, D.E. Pufhal & A.W. Clifton 1996. Model for the prediction of shear strength with respect to soil suction. *Can Geotech. J.* 33: 379-392.

Vaunat J., C. Jommi & A. Gens 1997. A strategy for numerical analysis of the transition between saturated and unsaturated flow conditions. *Proc. 6th Int. Symp. on Numerical Models in Geomechanics, Montreal, 2-4 July*: 297-302 Rotterdam: Balkema.

Vaunat, J., E. Romero & C. Jommi 2000. An elastoplastic hydro-mechanical model for unsaturated soils. In *Experimental Evidence and Theoretical Approaches in Unsaturated Soils; Proc. of an International Workshop, Trento, 10-12 April 2000, this issue.*

Wheeler, S.J. & D. Karube 1996. Constitutive modelling. *Proc. 1st Int. Conf. On Unsaturated Soils, Paris, 6-8 September 1995*: 1323-1356 Rotterdam: Balkema.

Wheeler, S.J. & V. Sivakumar 1995. An elastoplastic critical state framework for unsaturated soil. *Géotechnique* 45(1):35-53.

153

Experimental Evidence and Theoretical Approaches in Unsaturated Soils, Tarantino & Mancuso (eds)
© *2000 Taylor & Francis, ISBN 90 5809 186 4*

Modelling the behaviour of unsaturated silt

F. Geiser, L. Laloui & L. Vulliet
Laboratoire de Mécanique des Sols, Ecole Polytechnique Fédérale de Lausanne, Switzerland

ABSTRACT: The main patterns of unsaturated fine soils are first described on the basis of literature results and tests carried out on a remoulded silt. After a short introduction on existing elastoplastic models dedicated to unsaturated soils, a new model, δ_{1_unsat} is proposed. This model makes use of two independent stress variables: the "saturated effective stress" σ-u_w and the suction s. δ_{1_unsat} is an improvement and extension of an existing model originally developed for saturated soils (HISS-δ_1). It is able to reproduce the principal features of unsaturated soils: (a) existence on a drying path of a saturated domain with non-zero suction where plastic (irrecoverable) strains appear, (b) elastic behaviour on a drying path when the suction is greater than the air entry value, (c) wetting-induced collapse, (d) influence of mechanical stress state on the hydric behaviour, (e) increase of preconsolidation pressure with suction, (f) evolution of ultimate strength with suction. An extension to undrained conditions is also presented. The coupled hydro-mechanical formulation is deduced from the application of the theory of continuum mechanics to porous media. Finally, the model is modified in the framework of the disturbed state concept (DSC). This extension of δ_{1_unsat} permits the prediction of the important softening behaviour commonly observed for unsaturated soils under deviatoric loading.

1 INTRODUCTION

For many years, research work in Soil Mechanics has developed essentially through the study of two limit states: dry and saturated soils. However, a whole class of geotechnical problems is conditioned by the behaviour of soils in an intermediate state termed the unsaturated state. This is defined here as the state in which the soil is subjected to a negative pore-water pressure, if the pore-air pressure is equal to atmospheric pressure (respecting the traditional soil mechanics sign convention: compression being positive).

This unsaturated state can be observed in the construction of embankments and tunnels, in road foundations (compaction effects), near the ground surface (infiltration and evaporation phenomena in the surface layer above the water table) and in numerous environmental engineering problems. The understanding of the behaviour of unsaturated soils is, therefore, important for the design and analysis of geostructures.

Over the last decades, many experimental investigations have led to an improved understanding of the behaviour of unsaturated soils. Some patterns have now been well established. Although more experimental tests need to be carried out, it is now possible to develop rather complete constitutive models.

In this paper, a review of the main characteristics of unsaturated soils is proposed. The existing models are then discussed and a new constitutive law called δ_{1_unsat} is introduced and validated. Using a hydro-mechanical formulation, the coupling effect between generated pore-water pressure and soil volume variation are included in the δ_{1_unsat} model in order to predict the behaviour of water undrained tests. Additionally, the use of a damage concept is proposed to simulate softening for low degrees of saturation and low net mean pressures in the post-peak phase. Finally, the limitations and capacities of the model are discussed.

2 REVIEW OF THE MAIN CHARACTERISTICS OF UNSATURATED SOILS

Extensive experimental tests on unsaturated, remoulded silt from the region of Sion (Switzerland) are used for this study (Laloui et al. 1997, Geiser 1999). The main features of unsaturated soils are deduced from this experimental database and also from other results in the literature. Two types of behaviour may be distinguished:
- the hydric behaviour, which involves changes in suction;
- the mechanical behaviour, which corresponds to changes in external load.

2.1 Hydric behaviour

The hydric behaviour can be investigated by subjecting samples to drying and wetting cycles (Fig. 1a). Experimental results have, for example, been reported by Blight (1966), Fleureau et al. (1993) and Geiser (1999). Five main patterns can be observed.
(A1) On a drying path, the samples remain saturated until a suction s_e (called air entry suction) is reached. Beyond this particular point, the degree of saturation of the specimen decreases (Fig. 1c).
(A2) A first increase in suction induces irreversible deformation until a suction close to s_e is reached. Beyond this point, the behaviour becomes quasi-reversible (Fig. 1b).
(A3) A hysteretic behaviour is observed on drying-wetting cycles (Fig. 1b-c).
(A4) For some soils on a wetting path, a collapse phenomenon is observed (e.g. Jennings and Burland 1962, Matyas and Radhakrishna 1968)
(A5) The mechanical stress level influences the hydric behaviour. Globally, the suction-induced strains decrease with increasing mechanical stress and the air entry value increases slightly (Vicol 1990). Those observations are based on few experimental data and need to be confirmed.

2.2 Mechanical behaviour

To describe the mechanical behaviour of unsaturated soils correctly, it is widely accepted that the total stress σ_{ij}, the pore-air pressure u_a and the pore water pressure u_w can be combined into two independent stress tensors (Fredlund & Morgenstern 1977). In the literature, the net stress $\sigma_{ij}^* = \sigma_{ij} - u_a \delta_{ij}$ and suction $s = u_a - u_w$ are commonly chosen. However, it is also possible to analyse the experimental results with another combination, namely the "saturated effective stress" $\sigma'_{ij} = \sigma_{ij} - u_w \delta_{ij}$ and the suction s. The features of unsaturated soils are discussed with both approaches in the following sections.

2.2.1 Isotropic path
The principal characteristics of unsaturated soils observed on isotropic paths are the following (e.g. Matyas & Radhakrishna 1968, Alonso et al. 1990, Maâtouk et al. 1995, Sivakumar 1993, Geiser 1999).

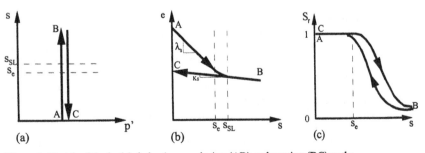

Figure 1. Sketch of the hydric behaviour on drying (AB) and wetting (BC) paths.

(B1) The preconsolidation pressure increases as the suction increases. This is observed in both stress combinations (σ^* and σ') for all soils.

(B2) Suction-induced stiffening effects can be seen when using a net mean stress interpretation. However, in the "saturated effective stress" approach, no clear trend has been observed on the basis of the existing experimental results, when using a semi-logarithmic scale (log (p')-e). In a linear scale representation (p'-e), a global stiffening of the soil with increasing suction is observed.

Based on few results, it is believed that the stiffening (B2) of the soil only starts for suctions greater than the air entry value. However, only a limited number of experimental results exist at small suction levels and the air entry value has often not been evaluated to confirm these assumptions. Similarly, in a net stress interpretation, the continuous increase of the preconsolidation pressure (B1) for small suction levels has not yet been proven.

2.2.2 Deviatoric paths

On deviatoric paths, the following patterns were observed for tests carried out at different suction levels.

(C1) In a net stress interpretation, the peak strength increased with suction (Delage & Graham 1995, Geiser 1999). This may be interpreted in the sense of a cohesion increase, while maintaining the friction angle constant.

In the "saturated effective stress" approach, a low increase in peak strength was observed for the Sion silt. The friction angle can also be considered to be constant. This remains to be confirmed for other soils.

(C2) At a constant initial net mean pressure, volumetric deformation on deviatoric paths of partially saturated soils decreased slightly as suction increased (Cui & Delage 1996). However, some authors have observed variable trends (Maâtouk et al. 1995, Sivakumar 1993).

In the saturated effective approach, it was observed for the Sion silt that volumetric strain decreased slightly at constant effective confinement pressure as suction increased.

(C3) Suction contributed to a small increase of the elastic rigidity of the soils for both stress combinations.

(C4) For high suction or small net mean pressure, an important loss of strength (brittle failure) was observed.

(C5) A unique critical state line can be observed in the stress plane (p^*-q or p'-q) for any value of suction.

3 EXISTING MODELS

In parallel with the experimental research work, several elastoplastic constitutive models were developed for unsaturated soils. Alonso et al. (1990) first proposed a critical state framework involving two independent sets of stress variables, namely the net stress and the suction. This approach was completed by the results of Wheeler & Sivakumar (1995). Fundamentally very similar to the previous model, the approach extends the concept of critical state to unsaturated soils adding one more state parameter, the water content. Wheeler (1996) modified the model by extending it to include relationships describing the variation of specific water volume (the volume of water and solids in a volume of soil containing unit volume of solids). Simultaneously, other models based on single effective stress were also developed. Their main advantage, when compared to models involving the net stress as a state variable, is a straightforward transition from the saturated to the unsaturated state (Kogho et al. 1993, Abou-Bekr 1995).

It is not our purpose here to describe the differences, advantages and limitations of all those models. As the Barcelona Basic Model (Alonso et al. 1990) is considered to be the most representative in elastoplastic modelling of unsaturated soils (Shen 1998, 2nd Int. Conf. On Unsaturated Soils, keynote lecture) and is also the one most currently used for further developments, it seems of interest to examine its capacity in predicting the previously-described features of unsaturated soils.

Tests carried out on the Sion silt were used to re-examine the Barcelona model capacity (Geiser 1999, Geiser et al. 1997a). Qualitatively as well as quantitatively, the predictions are good for mechanical paths. The isotropic behaviour is well predicted with increase in preconsolidation pressure (B1) and decrease of compressibility (B2) with suction. For deviatoric paths at constant net mean pressure, the model can predict the strength increase (C1-C5) and a small decrease in volume change (C2) with suction for normally consolidated soils. The collapse phenomenon can also be predicted (A4). However, some limitations are observed, mainly on hydric paths. The existence of a saturated domain with non-zero suction is ignored (A1) and the model cannot predict the first irreversible deformations on a drying path (A2). Another main problem induced by the choice of the net mean pressure as one of the stress variables is that the transition from the saturated to the unsaturated state and vice-versa is not straightforward.

To avoid this last problem and, more particularly, to improve the hydric predictions, a new model, $\delta_{1\text{-unsat}}$ was proposed by Geiser et al. (1999), based on another combination of stress variables: the saturated effective stress σ-u_w and the suction s. This model is able to reproduce the same patterns as the Barcelona model on mechanical paths. The hydric part is, however, very different and strongly influenced by the air entry value. Additionally, an extension of the model to damage concept is also proposed to predict the post-peak loss of strength observed for unsaturated soils when suction is large or the net mean pressure small (C4). The hydric hysteresis (A3) is for the time is ignored, as in most existing models.

4 DESCRIPTION OF THE $\delta_{1\text{-unsat}}$ MODEL

The hierarchical single surface model HISS-δ_1, developed initially for saturated soils (e.g. Desai 1994), has been modified to integrate the suction effect. The new model called $\delta_{1\text{-unsat}}$ was formulated within the framework of hardening plasticity using two independent sets of stress variables: excess of total mean stress over water pressure and suction. The detailed description of the model is given in Geiser (1999).

The increment of strain is decomposed into an elastic $\dot{\varepsilon}_{ij}^e$ and a plastic $\dot{\varepsilon}_{ij}^p$ part:

$$\dot{\varepsilon}_{ij} = \dot{\varepsilon}_{ij}^e + \dot{\varepsilon}_{ij}^p \tag{1}$$

The elastic increment is assumed to be composed of a mechanical $\dot{\varepsilon}_{ij}^{e^m}$ and a hydric $\frac{1}{3}\dot{\varepsilon}_v^{e^h}$ strain increment:

$$\dot{\varepsilon}_{ij}^e = \dot{\varepsilon}_{ij}^{e^m} + \frac{1}{3}\dot{\varepsilon}_v^{e^h}\delta_{ij} = \mathbf{D}_{ijkl}^{e-1}\dot{\sigma}_{kl}' + \frac{1}{3}A^{-1}\dot{s}\delta_{ij} \tag{2}$$

where $\mathbf{D}^e$ is the mechanical elastic tensor and A an elastic proportionality coefficient that describes the hydric behaviour. Despite the fact that experimental results have shown that the initial rigidity (correlated to Young's modulus) increases slightly with suction, the elastic tensor $\mathbf{D}^e$ is assumed here to be independent of suction and mechanical stress. The elastic hydric proportionality coefficient A also remains constant, as experimental results show no significant evolution with suction and mechanical stress (e.g. Biarez et al. 1993; Vicol 1990).

For writing the plastic strain increment, two yield surfaces and two plastic potentials (functions of suction s, stress σ' and the hardening parameter ξ) are defined:

$$F_1 = \tilde{F}_1(\sigma_{ij}',s,\xi) \qquad\qquad\qquad \text{mechanical yield surface}$$

$$F_2 = \tilde{F}_2(\sigma_{ij}',s,\xi) \qquad\qquad\qquad \text{hydric yield surface}$$

$$Q_1 = \tilde{Q}_1(\sigma_{ij}',s,\xi) \qquad\qquad\qquad \text{mechanical plastic potential}$$

$$Q_2 = \tilde{Q}_2(\sigma_{ij}',s,\xi) \qquad\qquad\qquad \text{hydric plastic potential}$$

The plastic strain increment is postulated to be composed of a mechanical and a hydric strain increment:

$$\dot{\varepsilon}_{ij}^{p} = \dot{\varepsilon}_{ij}^{p^m} + \frac{1}{3}\dot{\varepsilon}_{v}^{p^h}\delta_{ij} \tag{3}$$

where $\dot{\varepsilon}_{ij}^{p^m}$ is the mechanical plastic strain increment, associated to F_1 and Q_1 by :

$$\begin{cases} \dot{\varepsilon}_{ij}^{p^m} = \lambda_1 \partial_{\sigma_{ij}} Q_1 & \text{if } F_1 = 0 \quad and \quad if \quad \dfrac{\partial F_1}{\partial \sigma_{ij}} > 0 \\ = 0 & \text{if } F_1 < 0 \end{cases} \tag{4a}$$

In Equation (3), $\frac{1}{3}\dot{\varepsilon}_{v}^{p^h}$ is the hydric plastic strain increment, associated to F_2 and Q_2 by :

$$\begin{cases} \dot{\varepsilon}_{v}^{p^h} = 3\lambda_2 \partial_s Q_2 & \text{if } F_2 = 0 \quad and \quad if \quad \dfrac{\partial F_2}{\partial s} > 0 \\ = 0 & \text{if } F_2 < 0 \end{cases} \tag{4b}$$

The partial derivatives of the plastic potentials Q_1 and Q_2 give the direction of the strain increments, and λ_1 and λ_2 are the plastic multipliers. Their expressions, when changing either the external load at constant suction or the suction at constant external load, are:

$$\lambda_1 = \frac{\partial_{\sigma_{ij}} F_1 \mathbf{D}_{ijkl}^e \dot{\varepsilon}_{klm}}{\partial_{\sigma_{ij}} F_1 \mathbf{D}_{ijkl}^e \partial_{\sigma_{kl}} Q_1 - (\partial_\xi F_1)\gamma_{F1}} \quad \text{with } \gamma_{F1} = \left| \partial_{\sigma_{ij}} Q_1 \right| \tag{5a}$$

$$\lambda_2 = \frac{\partial_s F_2 A \dot{\varepsilon}_{vh}}{(\partial_s F_2) A (\partial_s Q_2) - (\partial_\xi F_2)\gamma_{F2}} \quad \text{with } \gamma_{F2} = \left| \partial_s Q_2 \right| \tag{5b}$$

4.1 Yield surfaces

Two yield surfaces were defined above. F_1 is derived from the saturated HISS-δ_1 model. It describes the soil yield in the p'-q plane at constant suction.

$$F_1 \equiv \frac{J_{2D}}{p_a^2} - \left[-\alpha(s) \left(\frac{J_1' + R(s)}{p_a} \right)^n + \gamma \left(\frac{J_1' + R(s)}{p_a} \right)^2 \right] F_s \tag{6}$$

where $F_s = 1 - \beta \overline{S}_r$

Expressions in Equation 6 are the followings:

J_{2D} is the second invariant of the deviatoric stress tensor, t_{ij}; J_1' is the first invariant of the "saturated effective" stress tensor $J_1' = 3p' = 3(p-u_w)$ and R is a bonding stress; p_a is a constant atmospheric pressure; γ and β are ultimate state parameters; $\overline{S}_r$ is the stress ratio with $\overline{S}_r = \sqrt{27/2} \, J_{3D} \cdot J_{2D}^{-3/2}$, with J_{3D} being the third invariant of the deviatoric stress tensor t_{ij}; α is the hardening function defined as:

$$\alpha(s) = \frac{a_1(s)}{\xi^{\eta_1}} \tag{7}$$

where a_1 and η_1 are the hardening parameters and ξ is the trajectory of total plastic strains given by $\xi = \int (d\varepsilon_{ij}^p d\varepsilon_{ij}^p)^{1/2}$; n is the phase change parameter related to the state of stress at which transition from compaction to dilation occurs or at which the change in the volume vanishes.

The experimental results on the Sion silt and those from the literature led to the development of two parameters at the suction level. As long as the soil behaves like a saturated soil ($s<s_e$), there is no reason to change the parameters with suction and they are kept constant:

- the hardening parameter $a_1(s)$ from Eq. 7 allows the expansion of the yield surface F_1 with suction (see Fig. 2, area 2):

$$\begin{cases} a_1(s) = a_1(0) = const. & if \; s < s_e \\ a_1(s) = a_1(0)[0.9\exp[-a_2[s-s_e]]+0.1] & if \; s \ge s_e \end{cases} \qquad (8)$$

where a_2 is a material parameter and s_e the air entry suction. It is of interest to note that the expansion of F_1 is not related to plastic strain generation. The choice of the coefficients 0.9 and 0.1 in Equation 8 was motivated by observations that the elastic compressibility of a soil is about $1/10^{th}$ of the plastic compressibility. This ratio has to be verified for other soils.

- the bonding stress $R(s)$ increases with the suction. This is in accordance with the assumption of a cohesion increase with suction (see Fig. 2, area 3):

$$\begin{cases} R(s) = R(0) & for \; s < s_e \\ R(s) = R(0) + r\sqrt{s} & for \; s \ge s_e \end{cases} \qquad (9)$$

The plastic non-associative potential Q_1 is defined as:

$$Q_1 = \frac{J_{2D}}{p_a^2} - \left[-\alpha_Q(s)\left(\frac{J_1'+R(s)}{p_a}\right)^n + \gamma\left(\frac{J_1'+R(s)}{p_a}\right)^2 \right]F_s \qquad (10)$$

with $\alpha_Q = \alpha + \kappa(\alpha_0 - \alpha)(1-r_v)$ $\qquad (11)$

where $r_v = \xi_v/\xi$, ξ_v is the volumetric part of ξ and α_0 is the α value at the beginning of shear loading and requires the non-associative material parameter κ. As excessive dilatations were predicted when assuming this parameter to be constant, a new relation was proposed allowing κ to evolve during the test:

$$\kappa = \kappa_\infty + \frac{\kappa_0 - \kappa_\infty}{\alpha_0}\alpha \qquad (12)$$

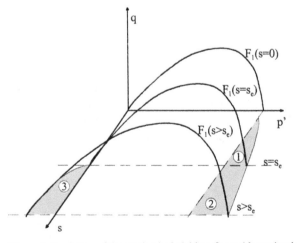

Figure 2. Evolution of the mechanical yield surface with suction in the p'-q-s plane.

with κ_0 being the non-associative parameter at the beginning of the test and κ_∞ corresponding to the final slope in the ε_1-ε_v plane.

The yield surface F_2 permits the description of hydric behaviour in the p'-s plane at constant "saturated effective" mean pressure. It was observed experimentally that the hydric behaviour changes when the suction is close to the shrinkage suction s_{SL} (Fig. 1b). As it was observed that s_{SL} is close to the air entry suction s_e (which is already used in the model as a material parameter), it will be assumed here that $s_{SL} \equiv s_e$. The mathematical expression of F_2 is similar to F_1:

$$F_2 \equiv -\left[-\alpha_h \left(\frac{3s}{p_a} \right)^n + \gamma \left(\frac{3s}{p_a} \right)^2 \right] F_s \qquad if \quad s < s_e \tag{13}$$

where $\alpha_h = \dfrac{a_3}{\xi^n}$, with a_3 being a hydric hardening parameter, is a hardening function controlling the elastic domain expansion for suction values below the air entry suction.

To model the reversible hydric behaviour beyond the air entry suction, the following condition is imposed:

$$F_2 < 0 \text{ when } s \geq s_e \tag{14}$$

The plastic potential Q_2 is simply defined as $Q_2 \equiv F_2$, assuming associative plasticity for hydric loading.

The yield surfaces F_1 and F_2 are coupled through the hardening variable ξ, with the increment of trajectory of total plastic strains $\dot{\xi}$ (see Fig. 2, area 1) defined as:

$$\dot{\xi} = (\dot{\varepsilon}_{ij}^{p^m} \dot{\varepsilon}_{ij}^{p^m})^{1/2} + \dot{\varepsilon}_v^{p^h} \tag{15}$$

This means that the plastic strain generated on a hydric path will influence the hardening of F_1. Similarly, the initial mechanical stress level influences the generation of hydric plastic strain, partly incorporating the characteristics described in (A5).

The yield surface evolution (Fig. 2) can be summarised as follow. Area 1 is due to volumetric strain hardening. It corresponds to an increase in size of the yield locus for normally consolidated soils and for constant saturated effective stress paths with increase of suction. This evolution will occur till the suction reaches the air entry value. Then the yield limit will expend in the Areas 2 and 3, which are not produced by any kind of hardening. They represent the influence of suction on the mechanical characteristics (Eqs. 8 & 9).

4.2 Summary

Finally, the proposed $\delta_{1\text{-unsat}}$ model contains 15 parameters of which only 5 are specific to unsaturated behaviour. Ideally, the determination of those 5 parameters requires:

- one drying and wetting test at constant p' to determine A, a_3 and s_e;
- three isotropic tests at different suction levels to determine the evolution of the preconsolidation pressure with suction (parameter a_2);
- three deviatoric tests at different suction levels to determine the evolution of the cohesion (parameter r).

4.3 Possible simplifications of the model

The model presented in the previous chapter is general and could be significantly simplified, reducing the unsaturated parameters to only two, namely a_2 and s_e.

As a first simplification, the bonding stress evolution parameter r could be neglected. It was indeed observed for other soils that at a same level of "saturated effective stress" the cohesion does not change significantly and can even slightly decrease (e.g. Trois-Rivières silt, Maâtouk

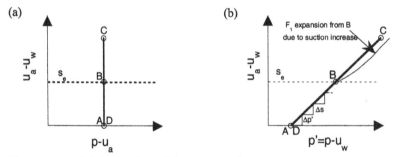

Figure 3. Typical drying-wetting path in a pressure plate.

et al. 1995; Kaolin, Sivakumar 1993). r could therefore be ignored without loosing much accuracy in the prediction.

The second simplification is related to the hydric behaviour. There are, to our knowledge, no useful results in the literature concerning suction loading under constant low saturated effective stress. Either the total pressure increases with the air pressure or the pore water pressure decreases, the first being typically observed in pressure plate tests and the second in hydric tests using the osmotic technique. The $\delta_{1\text{-unsat}}$ model is general and the introduction of a hydric yield surface F_2 enables the prediction of elastoplastic strains, assuming that the response of the soil would be similar to that on a drying path at constant net stress. However, it would be possible to predict the results of a standard drying test without F_2 and to assume that suction changes induce no volume changes at all at a constant p'.

Let us consider a pressure plate test with a first drying followed by wetting of the sample (Fig. 3a, path A-C-D). The pore-water pressure is constant and equal to the atmospheric pressure. The net mean pressure remains constant during the test and consequently the total pressure increases with pore-air pressure (Fig. 3a). Instead of increasing both s and p' simultaneously, small increments Δs and $\Delta p'$ are imposed for the purpose of modelling, where either p' or s are kept constant (Fig. 3b).

Below the air entry value (path A-B), Δs produces no deformation. However, the mechanical yield surface F_1 increases with the pressure $\Delta p'$, generating plastic deformations. Thus, the so-called "hydric slope" in the suction versus void ratio plane is identical to the saturated mechanical one. This has been observed by several authors (e.g. Zerhouni 1991, Fleureau et al. 1993, Taibi 1994).

Beyond the air entry value (path B-C), when the suction increases, F_1 expands without generating plastic strains as the hardening parameter a_1 is changing with s (Equation 8). When p' increases (at constant s), the soil begins inside the yield surface and elastic strains are generated. The experimental results on Sion silt show that the yield surface F_1 expands faster under a small step of suction Δs than it would for the same amount of pressure $\Delta p'$ (see expansion of F_1 on Fig. 2). Consequently, no further plastic strains are generated. When wetted (path C-D), the soil remains inside the yield surface and only elastic swelling strains are generated due to decreasing p'.

These assumptions need to be confirmed by showing experimentally that suction has no effect on volume when p' is constant. This would lead to major simplifications with only one yield surface remaining, and a reduction in the number of material parameters (A and a_3 would no longer exist).

4.4 Validation of $\delta_{1\text{-unsat}}$ in saturated conditions

The validation of the $\delta_{1\text{-unsat}}$ model is based on experimental results obtained for the Sion silt.

Table 1 shows the saturated parameters obtained for the Sion silt on the basis of one isotropic test and three deviatoric triaxial tests at different confining pressures. The predictions of the saturated tests are represented in the next section together with the unsaturated results.

4.5 Validation of $\delta_{1\text{-unsat}}$ in unsaturated conditions

4.5.1 Parameter determination

For the parameter determination, let us first consider the results obtained in a pressure plate with several samples. The volume and water content were determined on drying and wetting paths. The air entry value is estimated to be approximately 80 kPa. The elastic proportionality coefficient as is assumed to be equal to zero, following experimental observations of drying paths. The hydric hardening parameter a_3 is obtained by calibration.

The parameter a_2 allowing the expansion of F_1 beyond the air entry value is determined on the basis of the yielding points of three isotropic tests conducted at different suction levels (s=100, 200 & 280 kPa).

Finally, the bonding stress evolution parameter r is obtained by calibration with the observed evolution of the cohesion of the soil for the entire drained unsaturated triaxial tests.

The unsaturated parameters are listed in Table 2.

Table 1. Parameters of the $\delta_{1\text{-unsat}}$ model for the saturated state.

Symbol	Parameter	Value
E,	Elastic parameters	110 MPa,
ν		0.35
γ,	Ultimate parameters	0.0425,
β		0.58
a_1,	Hardening parameters	$2.58\ 10^{-7}$,
η_1		3.21
n	Phase change parameter	3.1
R	Bonding stress parameter	0
κ_0, κ_∞	Non-associative parameters	8, 0.42

Table 2. Parameters of the $\delta_{1\text{-unsat}}$ model for the unsaturated state.

Symbol	Parameter	Value
s_e	Air entry suction.	80 kPa
A	Elastic proportionality coefficient.	0 kPa
a_3	Hydric hardening parameter.	$7.5\ 10^{-6}$
a_2	Parameter for the evolution of parameter a_1 when suction $s>s_e$.	0.025
r	Evolution parameter for R	15 kPa

4.5.2 Simulation of drying and wetting paths

Figure 4 compares experimental results with the numerical simulation on hydric paths. These results have been used for the determination of s_e, A and a_3. Each experimental point corresponds to a different sample which was dried at constant net mean pressure in a pressure plate starting with a slurry (stress path see Fig. 3). Although the experimental scatter is significant, the model conforms to expectations with a first irreversible response, followed beyond s_e by a reversible one.

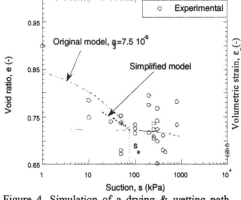

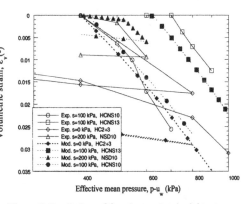

Figure 4. Simulation of a drying & wetting path, comparison with experimental results obtained in a pressure plate.

Figure 5. Prediction of four isotropic triaxial tests.

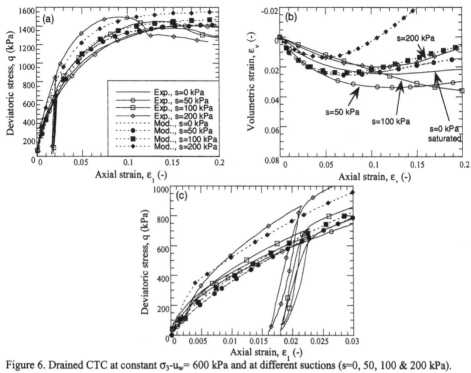

Figure 6. Drained CTC at constant σ_3-u_w= 600 kPa and at different suctions (s=0, 50, 100 & 200 kPa).

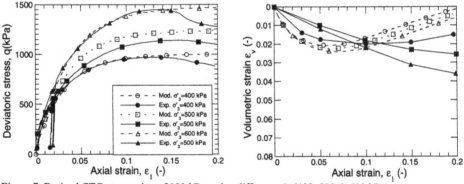

Figure 7. Drained CTC at a suction of 100 kPa and at different σ'_3 (400, 500 & 600 kPa).

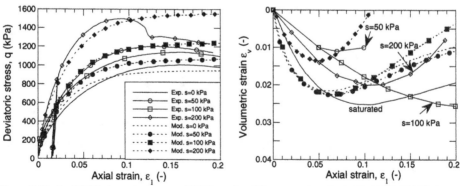

Figure 8. Drained CTC at constant σ_3-u_a= 400 kPa and at different suctions (s=0, 50, 100 & 200 kPa).

Unfortunately, when the soil is too soft, the determination of the sample volume is not accurate for small suctions (below 30 kPa for the Sion silt) with the method used . Thus, the initial irreversible slope is not well defined. The original model with two yield surfaces permits the prediction of any initial hydric slope (depending on the choice of the parameter a_3). However, the results are not in contradiction with the assumption that the hydric and mechanical slopes are identical. This corresponds to the simulation obtained with the simplified model represented in Figure 4. The reversible behaviour is similar in the two versions of the model, since in both cases the hydric proportionality constant is assumed to be zero. Thus, the elastic strains are only generated by the increase of the total pressure.

4.5.3 *Prediction of isotropic behaviour*
Figure 5 shows the simulation of four unsaturated, isotropic triaxial tests at different, constant levels of suction (s=0, 100 & 200 kPa). The saturated samples were first consolidated isotropically in a triaxial cell, then dried by imposing air pressure while keeping water pressure equal to atmospheric pressure until hydric equilibrium was reached. Finally, the total pressure p was increased by steps while maintaining constant air pressure. The predicted volumetric strains are in good agreement with the experimental results. The yield points are also reproduced satisfactory.

4.5.4 *Prediction of deviatoric behaviour*
Several triaxial shearing tests (CTC: conventional triaxial compression) were conducted on the Sion silt. The samples were prepared as described previously for the isotropic tests. After the drying process they were sheared in a triaxial cell in (water) drained conditions, while maintaining constant air pressure. The shear rates varied between 1.2 and 1.8 μm/min in order to allow dissipation of the excess pore-water pressure.

Figure 6 shows the simulation of four CTC triaxial tests at a confining effective pressure of 600 kPa and at different suction levels (s=0, 50, 100 & 200 kPa). The prediction of increasing peak strength with suction is good. The predicted curves at s=0 & 50 kPa are similar, as the bonding stress is assumed to be constant for suction under s_e and as no hydric plastic strains are generated at this pressure level during the drying process: this is in agreement with experimental observation. The brittle failure observed for s=200 kPa cannot be predicted with this formulation of the model. Figure 6b shows volumetric strain evolution vs. axial strain. Increasing dilatancy with suction is reproduced, and is in agreement with the experimental results, although exaggerated for larger suctions. This could be improved by introducing a dependency between the non-associative parameter $\kappa_{\_}$ and suction. Figure 6c is a zoom of Figure 6a, showing the capacity of the model for small deformation levels. The initial behaviour is well predicted.

Figure 7 shows the predictions of three triaxial CTC tests at a suction of 100 kPa and at different levels of confinement pressure (σ_3=400, 500 & 600 kPa). At this suction level, the predictions are good.

Figure 8 shows the simulation of four triaxial CTC tests at a confining net pressure of 400 kPa and at different suction levels (s=0, 50, 100 & 200 kPa). The test carried at s=50 kPa shows an atypical feature and should be considered with care. Conclusions similar to those for Figure 6 can be drawn.

4.5.5 *Collapse prediction*
The capacity of the model to predict collapse on a wetting path is tested. As no experimental results are available for the Sion silt (this soil seems not to be highly collapsible), a theoretical path is shown in Figure 9a. The samples were dried starting from a saturated slurry (A-B), p' was then increased while a constant suction (B-D) was maintained and, in some cases, p' was then reduced (D-D_1 or D-D_2). Finally, the samples were wetted at constant p' (D-E, D_1-E_1 or D_2-E_2). The evolution of the yield surface F_1 is also represented in Figure 9a.

Figure 9b-c shows predicted volumetric behaviour vs. p-u_w and s, respectively. For low levels of pressure (D_2-E_2), the sample only swells when being wetted, as the stress always remains inside the yield surface. For average p' (D_1-E_1), the sample first swells and then collapses, when reaching the yield surface. Plastic strains are generated, inducing hardening of F_1. Finally, for high p' (D-E), the sample volume only decreases (collapse) during the wetting process. Those

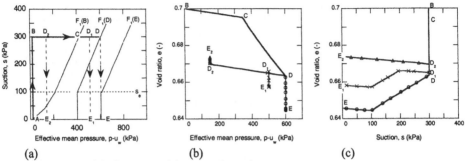

Figure 9. Response of the δ_l-unsat model on a wetting path.

predictions are in accordance with experimental observations: collapsible soils swell for low levels of pressure and collapse for higher pressures (Matyas & Radhakrishna 1968).

4.5.6 *Concluding remarks on the capacity of $\delta_{l\text{-unsat}}$*
The proposed model is able to reproduce most of the features of unsaturated soils on mechanical and hydric paths. The predictions are quantitatively good, except on deviatoric paths, where the predicted volume dilatancy is slightly excessive.

5 HYDRO-MECHANICAL COUPLING

5.1 *Pore water pressure determination*

To simulate the experimental behaviour during tests in which water cannot drain from the soil sample (which we define as an undrained condition) the coupling effect between the generated pore water pressure and the soil volume variation must be included in the model $\delta_{l\text{-unsat}}$ (Geiser et al. 1999, Geiser 1999). This is done using a hydro-mechanical formulation developed from the mixture theory by Hutter et al. (1999). Considering the fact that hydro-mechanical coupling exists as long as the water phase is continuous in the soil, the mass conservation equations for the water and solid constituents of the soil can be written:

$$\partial_t \rho_\alpha^* + div(\bar{\mathbf{v}}_\alpha)\rho_\alpha^* = 0 \tag{16}$$

where α represents the water ($\alpha = w$) or the solid ($\alpha = s$) constituent of the soil. ρ_α is the partial density defined as:

$$\rho_w^* = n S_r \rho_w \qquad \text{for water}$$

$$\rho_s^* = (1-n)\rho_s \qquad \text{for the solid skeleton;}$$

n is the porosity and S_r the degree of saturation. Equation 16 gives, for the water phase:

$$\frac{n}{\rho_w}\partial_t\rho_w + \frac{n}{S_r}\partial_t S_r + \partial_t n + n\,div(\mathbf{v}_w) = 0 \tag{17}$$

with $\dfrac{\bar{\partial}_t\rho_w}{\rho_w} = \beta_w \partial\mu_w$ and for the solid phase:

$$-\bar{\partial}_t n + (1-n)\beta_s\bar{\partial}_t u_w + (1-n)div\mathbf{v}_s = 0 \tag{18}$$

with $\dfrac{\bar{\partial}_t\rho_s}{\rho_s} = \beta_s\partial\mu_w$, where β_α is the compressibility of the phase α.

166

Adding Equations (17) and (18), neglecting the relative velocity of the water with respect to the solid skeleton and using the hypothesis of small deformations, one can obtain:

$$[n\beta_w + (1-n)\beta_s]\partial_t u_w + \frac{n}{S_r}\partial_t S_r + \partial_t \varepsilon_v = 0 \tag{19}$$

Then the pore-water pressure can be obtained as a function of the volumetric deformation and the variation of the degree of saturation as:

$$\partial_t u_w = -\frac{1}{[(1-n)\beta_s + n\beta_w]}(\partial_t \varepsilon_v + \frac{n}{S_r}\partial_t S_r) \tag{20}$$

5.2 Water compressibility

As shown in Equation 20, the pore-water pressure also depends on the water compressibility β_w (which includes the trapped air in the water for the unsaturated conditions). Many authors have tried to determine the compressibility of a mixture with water including trapped air. Generally, the compressibility is expressed considering the perfect gas laws for the air phase (Boyle) and the Henry law for the solubility of air. Neglecting the water compressibility, Bishop & Eldin (1960, quoted in Fredlund & Rahardjo 1993) proposed the following expression for the compressibility of the fluid mixture β_f

$$\beta_f = (1 - S_r + HS_r)u_{a0}/u_a^2 \tag{21}$$

where H is the Henry's constant, generally assumed to be 0.02; u_a is the average air pressure in the sample and u_{ao} the initial air pressure. For the experimental tests on the Sion silt used in the numerical simulations, which will be presented later, constant air pressure was imposed (i.e. $u_a = u_{a0}$).

Although the proposed approach is applied to a three-phase mixture, the determination of the pore-water pressure is limited to the use of the mass conservation equations of the solid phase and the water phase. That is why the Bishop & Eldin expression (21) was chosen for the evaluation of the compressibility of the water containing air bubbles ($\beta_f = \beta_w$).

5.3 Relation between degree of saturation and suction

To solve Equation 20, it is necessary to determine the dependence between the degree of saturation S_r and the suction s. Seker (1983) proposed the following water-saturation-suction relation, which was adopted for the $\delta_{1\text{-unsat}}$ model:

$$S_r = \frac{1}{1 + \left(\dfrac{\log(10 * s)}{\psi_0}\right)^{1/\psi_1}} \tag{22}$$

where ψ_0 and ψ_1 are material parameters and the suction s in kPa.

5.4 Water undrained tests

5.4.1 Parameter determination for the undrained tests
The Bishop and Eldin relation (Eq. 21) cannot be used for the saturated case ($u_a = u_{a0} = 0$). Fredlund and Rahardjo (1993) showed the effect of solubility of air in water on the compressibility of an air-water mixture (Fig. 10). The theoretical saturated compressibility value is $4.57 \ 10^{-7}$ kPa^{-1}. In our case, full saturation was never reached (max. $S_r \approx 96\%$) and, as a consequence, the compressibility β_w was taken for the saturated case as:

$$\beta_w = 2 \ 10^{-4} \text{ kPa}^{-1} \text{ (saturated case).}$$

This corresponds to the intersection between the curve including the solubility of air in water at atmospheric pressure of 101.3 kPa and a degree of saturation of 100% in Figure 10.

The Bishop and Eldin relation was applied for the unsaturated cases. The compressibility evolution was verified during the numerical simulations. For example, it was observed, with an air pressure equal to 100 kPa (corresponding to $S_{r\ initial} \approx 80\%$), at a confining pressure of 300 kPa, that the compressibility varied between:

$$\beta_w = 1.65\ 10^{-3}\ \&\ 1.02\ 10^{-3}\ kPa^{-1}\ (\text{unsaturated case}).$$

It is also necessary to determine (by calibration on a drying path) the parameters in the Seker relation (Eq. 22). Figure 11 shows good agreement between the experimental results and the Seker relation with the parameters $\psi_0 = 3.3$ and $\psi_1 = 0.055$.

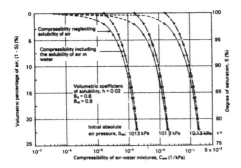

Figure 10: Effect of solubility of air in water on the compressibility of an air-water mixture (Fredlund & Rahardjo 1993).

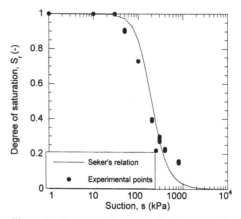

Figure 11. Relation s-S_r, comparison between the experience and Seker's relation.

5.4.2 Model predictions

Figure 12 shows the evolution of: (a) the deviatoric stress q, (b) the pore-water pressure u_w, and (c) suction s, versus axial strain ε_1 for constant water content tests carried out at a confining pressure of $\sigma_3 = 300$ kPa and air-pressure of $u_a = 0$, 100, 200 et 280 kPa. The prediction is good for the saturated reference case. The strength, water pressure and suction of the test run at $u_a = 100$ kPa are also well predicted. For larger air pressures and suctions, the samples fail brutally even before reaching the peak resistance (see Fig. 12a). It can be observed that the predicted water pressure generation reduces when the samples are drier, which is not really observed experimentally at this confinement pressure. This should be related to the early failure of the samples, which certainly disturbed not only the strength but also the pore pressure generation.

Figure 13 shows the results of the prediction of constant water content tests carried out at $\sigma_3 = 1000$ kPa and $u_a = 0$, 100, 200 and 280 kPa. The predictions are very good for the saturated case ($u_a = 0$) and when air pressure equals 100 kPa. The results are quantitatively in accordance with the experimental observations in all planes. For larger suctions, a brittle failure is again observed: the samples did not reach the peak strength. However, the disturbance appears later compared to the previous case (Fig. 12). It can be observed that the predicted water pressure evolution for higher suction is underestimated, but qualitatively in agreement with the predictions.

5.5 Conclusion concerning the hydro-mechanical coupling

It was shown for tests done on a silt in water undrained conditions that good predictions of pore-water pressure and strength are possible with the proposed $\delta_{1\text{-unsat}}$ model, providing Seker (1983) and Bishop & Eldin (1960) relations are used.

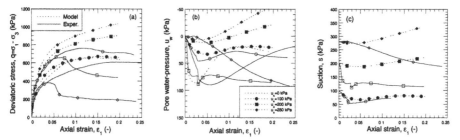

Figure 12: Comparison between the experimental results and the prediction of the model at σ_3=300 kPa and u_a =0, 100,200 & 280 kPa.

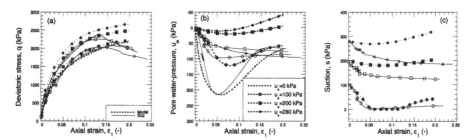

Figure 13: Comparison between the experimental results and the prediction of the model at $\sigma3$=1000 kPa and u_a =0, 100,200 & 280kPa.

6 EXTENSION OF THE MODEL WITH DSC

6.1 Introduction to DSC

The post-peak behaviour was proposed to be modelled in the general framework of the disturbed state concept (DSC, e.g. Desai 1995). The DSC is based on the idea that a deforming material element can be treated as a mixture of two constituent parts in the relative intact (RI) and fully adjusted (FA) states, referred to as reference states. During external loading, the material experiences internal changes in its microstructure due to a self-adjustment process and, as a consequence, the initial RI state transforms continuously to the FA state. The observed mean stress σ_{ij} is defined as:

$$\sigma_{ij} = (1 - D)\sigma_{ij}^i + D\,\sigma_{ij}^c \tag{23}$$

where σ_{ij}^i is the RI stress, σ_{ij}^c the FA stress and D the disturbance function. D varies between 0 (intact state) and 1 (adjusted state).

6.2 Boundary states

The DSC approach allows different choices for the description of the FA and RI states. The RI state can be represented by using the theory of elasticity, plasticity, viscoplasticity, etc. (Desai et al. 1996, Geiser et al. 1997b, Geiser 1999). Numerical tests have shown that assuming an elastic RI state leads to problems in the formulation, as the complete non-linear behaviour has to be included in the disturbance function. Consequently, the following approach was chosen (Fig. 14):

- the intact state is represented by the previously described $\delta_{1\text{-unsat}}$ model
- the fully adjusted state corresponds to a saturated state with a modified initial effective stress $p_{o}{}'_{\text{mod.}}$:

$$p_{o}{}'_{\text{mod.}} = (p_0{}' - u_a) + S_r (u_a - u_w) \tag{24}$$

The observation of the experimental tests with brittle failure on the Sion silt led to this modified stress. It was shown that, after failure, the soil tends to a smaller ultimate state than for an intact soil.

6.3 Disturbance function

The disturbance function D is expressed here as a scalar, which represents the microstructural changes leading to microcracking, damage and softening. It is proposed as a first approximation to express D as a function of the deviatoric stress invariant, as brittle failure mainly appears in the deviatoric plane (ε_1-q). The disturbance function is then:

$$D = \frac{\sqrt{J_{2D}}^{\,i} - \sqrt{J_{2D}}}{\sqrt{J_{2D}}^{\,i} - \sqrt{J_{2D}}^{\,c}} \tag{25}$$

with the exponent "i" corresponding to the RI state and "c" corresponding to the FA state (see Fig. 14).

The disturbance is assumed to be a function of the trajectory of the deviatoric plastic strains ξ_D and suction s. It is expressed as:

$$D = D_u (1 - e^{-B \xi_D^Z}) \tag{26}$$

where D_u is the ultimate value of D (a material parameter), which may be assumed to be independent of suction; B and Z are material parameters, functions of suction. The trajectory of the deviatoric plastic strains is defined as:

$$\xi_D = \int \left(dE_{ij}^p \cdot dE_{ij}^p \right)^{1/2} \tag{27}$$

where $dE_{ij}^p = d\varepsilon_{ij}^p - \frac{1}{3} d\varepsilon_v^p \, \delta_{ij}$, with $d\varepsilon_{ij}^p$ the plastic strain rate and $d\varepsilon_v^p = tr\left(d\varepsilon_{ij}^p \right)$.

Finally, the extension of the model to DSC involves the determination of three new parameters B, Z and D_u. B and Z are assumed to evolve with suction.

6.4 Validity of the DSC for post-peak failures

Shear bands were regularly observed on the Sion silt samples after the peak when conducting triaxial shear tests. It is possible that these bands were generated even before the peak was reached. As the post-peak strain is not homogenous anymore, the initialisation of the DSC,

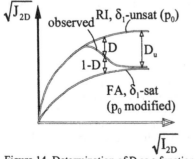

Figure 14. Determination of D as a function of $\sqrt{J_{2D}}$.

which assumes smeared cracking, would be open to criticism. Desai et al. (1997) has shown that in such cases the disturbance D would implicitly include a notion of characteristic length. This point has indeed to be studied more in depth.

6.5 Numerical simulations with the DSC

6.5.1 Determination of the parameters B and Z
The determination of D versus ξ_D and s is based on three unsaturated drained triaxial shear tests at a mean effective confining stress of 600 kPa at different suction levels. Applying Equation 25 through the entire stress path, together with computing of the plastic strain increments (from the elasto-plastic decomposition of strain) gives the evolution of D with ξ_D at a given suction. From Equation 26, the parameters Z and B are determined as follows:

$$\ln B + Z \ln(\xi_D) = \ln\left[-\ln\left(\frac{D_u - D}{D_u}\right)\right] \equiv D^* \tag{28}$$

A plot of D^* vs $\ln(\xi_D)$ (Fig. 15) gives the parameter Z as the slope and the parameter B as the intercept of the regression line. The values of the parameters are given in Table 3 and their evolution in Figure 16 for the Sion silt.

6.5.2 Validation of the DSC
The DSC extension of the model was tested for three investigations carried out at an effective confining pressure of 600 kPa. Figure 17 shows the simulations conducted at three different suction levels, using the parameters listed in Table 2. The constitutive model coupled with the DSC is able to reproduce the loss of strength after the peak at the different stress levels (Fig. 17). The volumetric behaviour is also well reproduced.

This DSC extension of the model only slightly influences the low-suction case (s=100 kPa, Fig. 17 a), as almost no loss of strength is observed for small suction levels.

The test conducted at a low degree of saturation (Fig. 17 c, s=280 kPa) is unique, as the sample never reached the peak value obtained for other tests at the same effective confining pressure. For this high level of suction, the brittle failure in the sample starts almost at the beginning of the test. Despite this, the simulation is very good with particularly close volumetric strains.

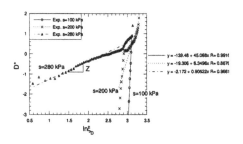
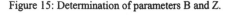
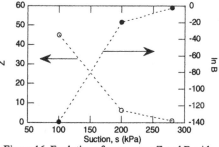

Figure 15: Determination of parameters B and Z.

Figure 16. Evolution of parameters Z and B with the suction.

Table 3: Parameters of the disturbance function (Eq. 25).

Suction s (kPa)	S_r^*	$p_{0\ mod}$	Z	B	D_u
100	0.85	584	45	$4.3\ 10^{-61}$	0.85
200	0.49	498	6.35	$4.1\ 10^{-9}$	0.85
280	0.31	407	0.91	0.114	0.85

$^*S_r(s)$ is determined on the basis of the Equation 22.

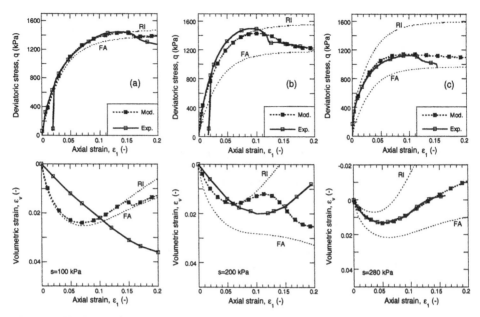

Figure 17. Simulation of triaxial shearing tests at a confinement pressure of σ_3'=600 kPa incorporating the DSC. (a) s=100 kPa, (b) s=200 kPa and (c) s=280 kPa.

6.6 Concluding remarks on the DSC extension

As it was shown, the prediction of the unsaturated soil behaviour is improved by the extension of the model to the DSC. Only three tests were used here and it is necessary to study the evolution of the disturbance function with the suction and the mechanical stress level further to obtain a complete model. However, it has been shown that the introduction of a damage concept permits the prediction of the brittleness typically observed for high levels of suction in shearing tests.

Finally, the DSC has indirectly permitted the introduction of a notion of critical state with the definition of the FA state. The expression proposed for the modified mean pressure is very close to expressions proposed for single effective stress in unsaturated soils (Geiser, 2000).

7 CONCLUSIONS AND PERSPECTIVES

A constitutive model, $\delta_{1\text{-unsat}}$, has been described by means of two independent stress variables, σ-u_w and s. Table 4 summarises the capacities of this model. Particularly, it has been shown that:

- The hydric behaviour is well predicted with $\delta_{1\text{-unsat}}$. A simplified form of the model (with no hydric yield surface F_2) has also proved to be efficient if the isotropic mechanical compressibility is taken equal to the hydric one for suction below the air entry value. Collapse under wetting can also be predicted by the model.
- The mechanical behaviour is qualitatively and quantitatively well predicted in the studied range of suctions and stresses. However, the volumetric strains are overestimated for large suction levels in triaxial shearing. An adaptation of the evolution of the non-associative parameter κ_{∞} with suction could correct this problem. It has also been observed that the introduction of a damage concept by means of the DSC improved this aspect.
- The incorporation of the DSC allows the prediction of brittle failures for high levels of suction. The basic HISS–δ_1 model (and thus $\delta_{1\text{-unsat}}$) is not based on a critical state concept. However, the extension of the model to the DSC allows the incorporation of this notion.

Table 4: Capacity of the model $\delta_{1\text{-unsat}}$.

Behaviour	Capacity of the model $\delta_{1\text{-unsat}}$	Qualit.	Quanti.	Improvement
HYDRIC				
A1. Saturated domain with non-zero suction	Is incorporated with the introduction of changes when $s>s_e$.	☺	☺	
A2. Irreversible strain below se on a first drying, followed by reversible behaviour	Good predictions. With the simplified model, the hydric compressibility must be identical to the mechanical one.	☺	☺	
A3. Hydric hysteresis	Is ignored in the model.	✘	✘	
A4. Collapse under wetting	Is incorporated in the model.	☺	?	
A5. Influence of the mechanical stress level on the hydric response	The strain generation is a function of the pressure level, as both yield surfaces are correlated through the trajectory increment of the total plastic strains.	☺	?	
B. MECHANICAL (isotropic)				
B1. Increase of the preconsolidation pressure	Is incorporated in the model.	☺	☺	
B2. Decrease of the compressibility (linear scale)	Is incorporated in the model.	☺	☺	
C. MECHANICAL (deviatoric)				
C1. Increase of the peak strength with s	Good prediction of the model.	☺	☺	Could be modified depending on different soils
C2. Volumetric behaviour	Predictions are in agreement with observation, but with an overestimation of the dilatancy for high suctions.	☺	☺	Better with DSC or with $\kappa_\infty = \kappa_\infty(s)$
C3. Brittle failure for high level of s	Not predicted.	✘	✘	Included in DSC ☺
C4. Increase of elastic rigidity with suction	Is ignored, but it does not strongly influence the results.	✘	✘	
C5. Critical state	δ_1-unsat is not a critical state based model.	✘	✘	Can be incorporated with DSC

☺= good, ☺ = satisfactory, ⊗ = bad et ✘ = not predicted by the model

The proposed expression of the critical state (= Fully Adjusted state) is close to a single effective stress approach in unsaturated soils and gives quantitatively good results.

It was also shown for tests done on a silt in water undrained conditions, that good predictions of pore water pressure and strength are possible with a hydro-mechanical formulation added to the δ_1-unsat model, coupled with a suction-degree of saturation relation (Seker 1983) and a water compressibility evolution with suction (Bishop & Eldin 1950).

Finally, $\delta_{1\text{-unsat}}$ requires 15 parameters from which five relate to unsaturated aspects: three for the hydric behaviour, one for the isotropic mechanical behaviour and one enabling the evolution of the peak strength. The simplified version of $\delta_{1\text{-unsat}}$ only requires two unsaturated parameters and still results in good predictions. As a saturated effective stress $\sigma\text{-}u_w$ was adopted as a state variable in this constitutive model, the transition from saturated state to unsaturated state is straightforward.

ACKNOWLEDGEMENTS

This work was supported by the Swiss NSF.

REFERENCES

Abou-Bekr, N. 1995. *Modélisation du comportement mécanique et hydraulique des sols partiellement saturés*. Doctoral thesis, Ecole Centrale Paris.

Alonso, E.E., A. Gens & A. Josa 1990. A constitutive model for partially saturated soils. *Géotechnique 40(3)*: 405-430.

Biarez, J., J. M. Fleureau & S. Taibi 1993. Constitutive model for unsaturated granular media. *Powders and Grains*: 51-58.

Bishop, A.W. & G. Eldin 1950. Undrained triaxial tests on saturated sands and their significance in the general theory of shear strength. *Géotechnique 2(1)*: 13-32.

Blight 1966. Strength characteristics of desiccated clays. *ASCE J. Soil Mech. Found. Eng. Div., 93*: 125-149.

Cui, Y.J. & P. Delage 1996. Yielding and plastic behaviour of an unsaturated compacted silt. *Géotechnique 46(2)*: 291-311.

Delage, P. & J. Graham 1995. Mechanical behaviour of unsaturated soils: Understanding the behaviour of unsaturated soils requires conceptual models. *Proc. 1st Conf. On Unsaturated Soils, Paris*: 1223-1256. Rotterdam: Balkema.

Desai, C.S. 1994. Hierarchical single surface and the disturbed state constitutive models with emphasis on geotechnical applications. *Geotechnical engineering: Emerging trends in design and practice*: 115-154. Saxena Ed.

Desai, C.S. 1995. Constitutive modelling using the disturbed state as microstructure self adjustment concept. *Continuum models for materials with microstructures*. H.B. Mülhaus Ed., John Wiley & Sons.

Desai, C.S., L. Vulliet, L. Laloui & F. Geiser 1996. *Disturb state concept for constitutive modeling of partially saturated porous materials*. Internal report, EPFL.

Desai, C. S., C. Basaran & W. Zang 1997. Numerical algorithms and mesh dependence in the disturbed state concept. *Int. Jour for Num. Meth.in Eng., 40 (16)*: 3059-3083.

Fleureau, J.M., S. Kheirbek-Saoud, R. Soemitro & S. Taibi 1993. Behavior of clayed soils on drying-wetting paths. *Can. Geotech. J. 30*: 287-296

Fredlund, D.G & N.R. Morgenstern 1977. Stress state variables for unsaturated soils. *ASCE J. Geotechnical Eng. Div. 103(GT5)*: 447-465.

Fredlund, D. G. & H. Rahardjo 1993. *Soil mechanics for unsaturated soils*. John Wiley & Sons.

Geiser, F., L. Laloui & L. Vulliet 1997 (a). Constitutive modelling of unsaturated sandy silt. *Proc. Computer and Advances in Geomechanics, Wuhan*: 899-904.

Geiser F., L. Laloui, L. Vulliet, C.S. Desai 1997 (b). Disturbed state concept for partially saturated soils. *Numerical Models in Geomechanics, NUMOG VI, Montréal*: 129-133.

Geiser F., L. Laloui & L. Vulliet 1999. Unsaturated soil modelling with special emphasis on undrained conditions. *Numerical Models in Geomechanics, NUMOG VII, Graz*: 9-14.

Geiser, F. 1999. *Comportement mécanique d'un limon non saturé: étude expérimentale et modélisation constitutive*. PhD Thesis, Swiss Federal Institute of Technology, Lausanne.

Geiser, F. 2000. Applicability of a general effective concept to unsaturated soils. *Asian Conf. On Unsat. Soils, Singapore*

Jennings, J.E.B. & J.B. Burland 1962. Limitations to the use of effective stresses in partly saturated soils. *Géotechnique 12(2)*: 125-144.

Kogho, Y., M. Nakano & T. Myazaci 1993. Theoretical aspects of constitutive modelling for unsaturated soils. *Soils and Foundations 33(4)*: 49-63.

Hutter, K., L. Laloui & L. Vulliet 1999. Thermodynamically based mixture models of saturated and unsaturated soils. *Journal of mechanics of cohesive-frictional materials, 4(2)*: 295-338.

Laloui, L., F. Geiser, L. Vulliet, X.L. Li, R. Charlier. & A. Bolle 1997. Characterization of the mechanical behaviour of an unsaturated sandy silt. *Proc. 14th Conf. on Soil Mech. and Found. Eng., Hambourg* 347-350. Rotterdam: Balkema

Maâtouk, A., S. Leroueil & P. La Rochelle 1995. Yielding and critical state of collapsible unsaturated silty soil. *Géotechnique 45(3)*: 465-477.

Matyas, E. L. & H. S. Radhakrishna 1968. Volume change characteristics of partially saturated soils. *Géotechnique 18(4)*: 432-448.

Seker, E. 1983. *Etude de la déformation d'un massif de sol non saturé*. PhD Thesis, Swiss Federal Institute of Technology, Lausanne.

Shen, Z. 1998. Advances in numerical modeling of deformation behaviors of unsaturated soils. *2nd Int. Conf. On Unsaturated Soils, Beijing*: 180-193. Int. Acad. Publishers.

Sivakumar, V. 1993. *A critical state framework for unsaturated soils*. PhD thesis, University of Sheffield.

Taibi, S. 1994. *Comportement mécanique et hydraulique des sols partiellement saturés*. PhD Thesis, Ecole Centrale, Paris

Vicol, T. 1990. *Comportement hydraulique et mécanique d'un limon non saturé: application à la modélisation*. PhD Thesis, Ecole Nationale des Ponts et Chaussées, Paris.

Wheeler, S. J. & V. Sivakumar 1995. An elasto-plastic critical state framework for unsaturated soil. *Géotechnique 45(1)*: 5-53.

Wheeler, S. J. 1996. Inclusion of specific water volume within an elasto-plastic model for unsaturated soil. Canadian Geotechnical Journal 33: 42-57.

Zerhouni, M.I. 1991. *Rôle de la pression interstitielle négative dans le comportement des sols- Application aux routes*. PhD Thesis, Ecole Centrale, Paris.

APPENDIX: NOTATIONS AND DEFINITIONS

A	elastic hydric proportionality coefficient (δ_1-unsat)	p_a	atmospheric pressure
B	DSC material parameter	q	deviatoric stress, $\sigma_1 - \sigma_3$
D	disturbance function	r	suction induced evolution parameter for R in δ_1-unsat
D^e	mechanical elastic tensor	s	matrix suction , u_a-u_w
D_u	ultimate disturbance function	s_e	air entry suction
F_1	mechanical yield surface (p'-q plane)	s_{SL}	shrinkage suction
F_2	hydric yield surface (p'-s plane)	t_{ij}	deviatoric stress tensor
J_1	first stress invariant	u_a	pore-air pressure
J_{2D}	second invariant of the deviatoric stress t_{ij}	u_w	pore water pressure
J_{3D}	third invariant of the deviatoric stress t_{ij}	β_α	compressibility of phase α
Q_1	mechanical plastic potential (p'-q plane)	ε_1	axial strain
Q_2	hydric plastic potential (p'-s plane)	ε_v	total volumetric strain $\varepsilon_v = \Delta V/V$
R	bonding stress saturated parameter (δ_1-unsat)	η_1	mechanical hardening parameter in δ_1-unsat
S_r	degree of saturation	κ_0, κ_∞	non–associative parameters in δ_1-unsat
Z	DSC material parameter	ξ	trajectory of total plastic strain
a_1	mechanical hardening parameter (δ_1-unsat)	ξ_D, ξ_V	deviatoric, respectively volumetric part of ξ
a_2	suction induced evolution parameter for mechanical hardening in δ_1-unsat	ρ_α	density of phase α
a_3	hydric hardening parameter (δ_1-unsat)	σ_1	axial stress
e	void ratio	σ_3	radial or confinement stress
n	porosity	σ_{ij}	total stress
n	phase change parameter (δ_1-unsat)	σ_{ij}'	saturated effective stress $\sigma_{ij} - u_w$
p	total mean pressure, $p=\sigma_{kk}/3=(\sigma_1 +2 \sigma_3)/3$	σ_{ij}^*	net stress $\sigma_{ij} - u_a$

Experimental Evidence and Theoretical Approaches in Unsaturated Soils, Tarantino & Mancuso (eds)
© *2000 Taylor & Francis, ISBN 90 5809 186 4*

Interpretation of shear response upon wetting of natural unsaturated pyroclastic soils

M. V. Nicotera
Dipartimento di Ingegneria Geotecnica, Università degli Studi di Napoli Federico II, Italy

ABSTRACT: A theoretical interpretation of observed shear responses upon wetting of two unsaturated pyroclastic soils is reported. The experimental observations have been obtained by direct shear tests performed on specimens at natural water content and on saturated specimens. In some cases natural water content specimens have been wetted during shearing stage at peak strength. This procedure has been used to analyse all the phenomena that occur when complete saturation is performed in proximity of shear failure. In every test the wetting process produced either a reduction of shear strength either a modification of dilatancy. The observed behaviour has been interpreted in the framework of critical state soil mechanics as originally extended to unsaturated soils by Alonso et al. (1987) and, subsequently, improved by a number of other authors.

1 INTRODUCTION

The city of Naples (Italy) is located in the center of a wide volcanic region constituted by the *Somma Vesuvio* volcano rising few kilometres south east from the center of the city and the volcanic district of *Campi Flegrei* located a few kilometres west. The subsoil of the urban area mainly consists of pyroclastic materials (soils and rocks) produced by eruptions of the volcanic district of *Campi Flegrei*. Pyroclastic soils (pozzolana) generally overlay the lithic part of the formation (Neapolitan Yellow Tuff and other pyroclastic tuffs or soft rocks). In a large part of the city the pozzolana is located above the groundwater table in a partially saturated condition. Due to exceptional rainfalls or leaks from aqueducts and sewerage the saturation of this type of soils occasionally causes slides in natural and artificial slopes and settlements in other parts of subsoil.

Pyroclastic soils originated by eruptions of Campi Flegrei and Somma Vesuvio are even present as cover of the dolomitic limestone slope of Sorrentina Peninsula. In this area the shallow layers of pyroclastic deposits occasionally suffer landslides as flowslides. Those instability phenomena are triggered by meteoric events which produce large variations of the degree of saturation of the involved soils.

Although the connection between the occurrence of this kind of phenomena and the modification of the degree of saturation is known since long time, only after the recent development of a clear and comprehensive framework for the mechanics of partially saturated soils it is possible to approach these problems in a rigorous way. A research program on this topic started at the University of Naples a few years ago to define the main physical properties and the mechanical behaviour of this kind of unsaturated pyroclastic soils (Nicotera 1998, Aversa et al. 1998, Nicotera et al. 1999a, b)

In the framework of a large research project on unsaturated pyroclastic soils co-ordinated by prof. A. Evangelista, a laboratory for partially saturated soils was designed and assembled (Nicotera 1998, Nicotera & Aversa 1999). A wide laboratory investigation, involving pressure plate, oedometer, triaxial and direct shear tests was carried out. The aim of the research was to verify if the effects of the transition from unsaturated to saturated condition on the mechanical behaviour of the studied pyroclastic soils are so relevant to justify the observed instability phenomena. Therefore the research activity was focused to investigate the mechanical response of unsaturated pyroclastic natural soil upon wetting.

2 TESTED SOILS

The experimental program was carried out on two different pyroclastic soils. The first one is a pozzolana owing to the same formation of Neapolitan Yellow Tuff (Cole et al. 1993, Scarpati et al. 1993). This soil, in the paper indicated as soil A, is relatively homogeneous and widely spread out in the geographical area of Naples. It was recovered in cylindrical samples (83 mm in diameter and 300 mm in height) in the north-east part of the urban area of Naples at a depth varying between 12.5 m and 14.5 m, well above the groundwater table. The second soil, in the paper indicated as soil B, is a phlegrean pozzolana typically present at the bottom of the pyroclastic cover of the dolomitic limestone slopes in Sorrentina Peninsula (Scotto di Santolo et al. in press). It was sampled in different locations which has been affected by flowslides in the last years (Scotto di Santolo 2000). The main physical properties of the two soils are reported in Table1. Figure 1 shows that the two soils have a similar grains size distribution and are well graded.

3 OBSERVED BEHAVIOUR OF PYROCLASTIC SOILS UPON WETTING

3.1 Experimental program and techniques

The first step in the research program was the development of two suction controlled apparatuses: a triaxial cell and an oedometer. Both apparatuses use axis translation techniques to control suction. In the triaxial one it is possible to perform tests on unsaturated material following any triaxial stress-path controlling either net stress either matric suction. As previously pointed out one of the main scope of the research program was to analyse the mechanical response of studied material during suction reduction stage (wetting). Both volumetric and shear response of the ma-

Table 1. Mean values of main physical properties of tested soils in situ.

	γ (kN/m³)	γ_d (kN/m³)	e	w_0	S_{r0}	G_s
Soil A	13.14	10.57	1.271	0.246	0.473	2.43
Soil B	11.33	9.01	1.986	0.273	0.356	2.63

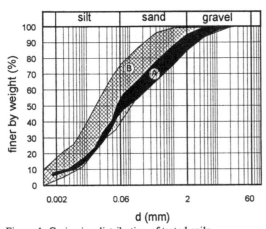

Figure 1. Grain size distribution of tested soils.

178

terial A have been experimentally investigated in this apparatus along wetting paths (Nicotera 1998).

The collapse of pozzolana was already observed some years ago performing conventional oedometer tests on natural sample (Pellegrino 1967); in those tests the unsaturated sample showed a marked collapsible behavior when submerged by water under different value of vertical net stress. Analogous results were obtained in suction controlled oedometric tests on material A. This collapsible behavior seems to be one of the key factors of some settlements phenomena which occur in these pyroclastic deposit.

Tests performed on material A showed that equalisation stage in triaxial apparatus were unacceptably time consuming (Nicotera 1998). This finding suggests the necessity to adopt a different experimental technique in order to observe mechanical response of pyroclastic material upon wetting. At this stage of the research program qualitative observation of mechanical behaviour was preferred instead of quantitative description. Thus, conventional direct shear tests were chosen although a triaxial stress-path and suction controlled apparatus was available. Direct shear apparatus, in fact, even less sophisticated of triaxial one, was judged more suitable to observe some particular phenomena, since it allows the almost complete saturation of the sample in very short time, without any interruption of the test.

Three different kinds of direct shear tests have been performed on both soil A and soil B: a) tests on natural water content samples; b) tests on samples saturated in the consolidation stage; c) test on natural water content samples wetted during shearing stage at peak strength. The last ones have been used to analyse all the phenomena that occur when complete saturation is performed in proximity of shear failure, in order to analyse shear resistance evolution during the process of suction zeroing. Direct shear tests on saturated samples have been performed for comparisons.

Direct shear tests on material A have been carried out at three different values of vertical stress: 19 kPa, 177 kPa, 345 kPa. These values corresponds respectively to the minimum allowed by the adopted shear apparatus, the in situ vertical stress and twice the in situ vertical stress; they have been chosen to analyse the effect of stress level on observed behaviour. Tests on material B have been carried out at vertical stresses ranging from 19 kPa to 54 kPa. The choice of lower stress level in the case of material B relays on the simple observation of its relevance in the analysis of instability phenomena affecting the pyroclastic cover of the dolomitic limestone slope of Sorrentina Peninsula.

3.2 Results of direct shear tests

The variation ranges of the physical properties of tested specimens are reported in Table 2. Typical results of tests on material A and material B are shown in Figures 2-4 and Figures 5-7.

Material A at natural water content shows a brittle behaviour when sheared at the lower stress level and at in situ stress; on the contrary material A shows an hardening behaviour either at the higher stress level either saturated (no matter of stress level). The wetting of material A in correspondence of peak strength (type c test) produces a sudden and relevant decrease of strength and an abrupt settlement. With increasing shearing, shear strength begin to increase reaching a stationary value; also vertical settlements became stationary. In the first part of shear tests, the samples initially subjected at lower values of vertical stresses tends to dilate; while the samples subjected to higher values of vertical stresses contract. After saturation, however, all the samples have a contracting behavior. Similar behaviour is shown by Material B.

Table 2. Physical properties of tested samples after consolidation stage, in parenthesis overall water content after wetting at peak strength.

	γ_d (kN/m^3)	e	w	S_r
Soil A: specimens at natural water content	8.56÷10.57	0.968÷1.783	0.173÷0.387 (0.414÷0.702)	0.272÷0.633
Soil B: specimens at natural water content	8.61÷9.54	1.700÷1.984	0.109÷0.199 (0.405÷0.687)	0.155÷0.263
Soil B: "saturated" specimens	8.59÷10.22	1.717÷2.004	0.533÷0.777	0.817÷1.000

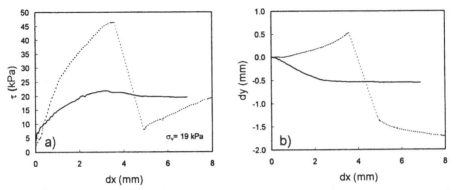

Figure 2. Typical results of direct shear tests on soil A (saturated specimen = continuous line, specimen wetted at peak = dotted line).

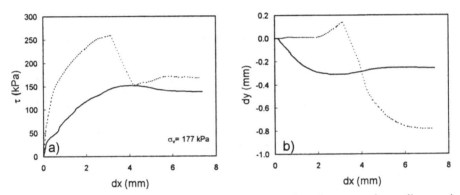

Figure 3. Typical results of direct shear tests on soil A (saturated specimen = continuous line, specimen wetted at peak = dotted line).

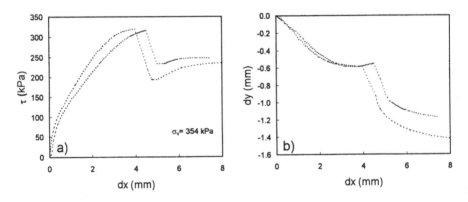

Figure 4. Typical results of direct shear tests on soil A (saturated specimen = continuous line, specimen wetted at peak = dotted line).

3.3 Strength envelopes

Figure 8 shows, in $\{\sigma_v, \tau\}$ plane, peak and final strengths of all shear tests performed on soil A: open circles represent peak strength of specimens tested at natural water content (type a tests);

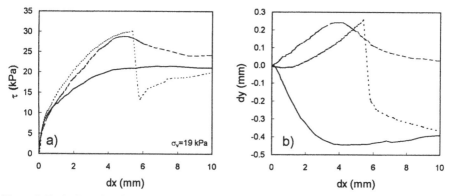

Figure 5. Typical results of direct shear tests on soil *B* (saturated specimen = continuous line, specimen wetted at peak = dotted line, specimen at natural water content = dashed line).

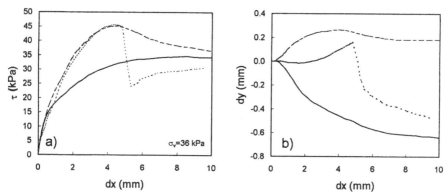

Figure 6. Typical results of direct shear tests on soil *B* (saturated specimen = continuous line, specimen wetted at peak = dotted line, specimen at natural water content = dashed line).

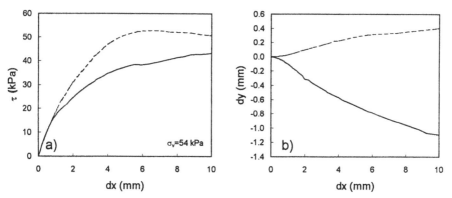

Figure 7. Typical results of direct shear tests on soil *B* (saturated specimen = continuous line, specimen at natural water content = dashed line).

filled circles represent final strength of both saturated specimens (type b tests) and wetted at peak specimens (type c tests). As it was expected specimens at natural water content show a significant higher strength than the saturated ones. However points representing peak strength of natural water content material are more disperse than the ones representing final strength of saturated material.

Diagrams in Figure 9a, b show, in $\{\sigma_v, \tau\}$ plane, peak and final strengths of all shear tests performed on soil B. In this case it can be noticed that final strength of material at natural water content is quite similar to final strength of saturated material.

The strength data have been interpreted in terms of net stress (i.e. total stress) to obtain the strength envelopes of the two material. In Table 3 friction angles and cohesive intercepts relative to each strength envelope are reported.

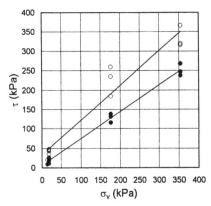

Figure 8. Strength envelopes of soil A: peak strength of specimens at natural water content (= open circles); final strength of saturated specimens and of specimens wetted at peak (= filled circles).

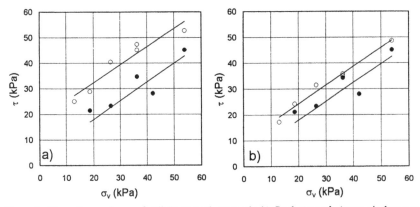

Figure 9. Strength envelopes of soil B: a) peak strength; b) final strength (open circles = specimens at natural water content; filled circles = saturated specimens).

Table 3. Strength parameters as inferred from experimental data

	Peak		Final	
	ϕ (°)	c (kPa)	ϕ (°)	c (kPa)
Soil A: natural water content	41.96	32.07		
Soil A: wetted at peak			34.63	6.07
Soil B: natural water content	35.35	18.09	36.01	9.64
Soil B: saturated	36.01	3.53	36.01	3.35

182

Cohesion intercept c of strength envelopes of natural water content material was obtained by means of a linear interpolation in $\{\sigma_v, \tau\}$ plane of the points representative of maximum strengths. The values of c (equal to 30 kPa for material A 18 kPa for material B) obtained in this way are influenced not only by the "natural" suction, but also by the approximation of strength envelope to a line.

In the case of material saturated during consolidation stage, peak and final strength envelopes differ very little one from the other. The value of cohesive intercept of peak envelope obtained from a linear interpolation in $\{\sigma_v, \tau\}$ plane, is probably influenced by the curvature of envelope itself. Final strength envelope approximates critical state line at zero suction.

4 THEORETICAL INTERPRETATION OF OBSERVED BEHAVIOUR

4.1 Theoretical framework

The observed behaviour can be interpreted in the framework of critical state soil mechanics as originally extended to unsaturated soils by Alonso et al. (1987) and, subsequently, improved by a number of other authors. In the paper, the model proposed by Wheeler & Sivakumar (1995) will be adopted.

The stress state of an unsaturated soil is represented in triaxial conditions by three stress variables: the mean net stress p ($= \Sigma \sigma_i / 3 - u_a$); the matric suction s ($= u_a - u_w$); the deviatoric stress q ($= \sigma_1 - \sigma_3$). Thus the stress state is represented by a point in the triaxial stress space $\{p, s, q\}$. A curve called the loading-collapse curve (LC) of equation $p = p_0(s)$ represents in the $\{p, s\}$ stress plain the yield locus in isotropic conditions, while a surface in the $\{p, s, q\}$ stress space defines the critical state conditions towards which every shearing process converges. The critical state surface (CSS) can be expressed by the equation:

$$q = M(s) \cdot p + \mu(s) \qquad (1)$$

The intersection of CSS with a generic plane at constant suction is represented by a line called the critical-state line (CSL). In the $\{p, q\}$ plane M and μ represent respectively the slope of CSL and his intersection with q axis. In their model Alonso et al. (1990) assumed a constant M value; this assumption would be equivalent to a constant value of ϕ' in the shear strength equation as proposed by Fredlund et al. (1978). The simple assumption of M = cost will be adopted in the following analysis.

The yield-surface (YS) in the $\{p, s, q\}$ stress space represents the boundary of the elastic domain. The intersection of YS with a generic plane at constant suction is assumed to be an ellipse (yielding curve YC) having axes parallel to those of deviatoric and mean stresses. The YC intersection with the positive p axis represents the yield stress p_0 in isotropic conditions and hence belongs to the LC; indeed the LC is the intersection of the YS with the isotropic plane. The ratio M^* between the ellipse's axes doesn't correspond to the inclination M of the CSL and changes with both suction and yielding stress p_0 in isotropic loading conditions.

The intersections of yielding and critical state surfaces with a generic plane at constant suction show the aspect reported in Figure 10, in which either the dimensions and the ratio between the ellipse's axes depends from suction. Every stress path inside the elastic domain produces recoverable strains; on the contrary when the point representative of the stress state approaches the YS and moves further in outward direction, it can causes an enlargement or a reduction of the domain itself. During the enlargement unrecoverable volumetric compressive strains are generated. Hardening is assumed to be correlated with plastic volumetric strains by means of a unique parameter, the mean yielding stress at zero suction $p_0(0)$. The flow rule is associated, and so the plastic strain increment vector ($\delta\varepsilon_v^p$, $\delta\varepsilon_s^p$) is in the direction of the outward normal to the yield surface in the $\{p, s\}$ stress plain. Wheeler & Sivakumar model is devoted only to the study of stress paths which intersect yielding ellipse in points that are located at the right side of the intersection between the ellipse and the critical state line.

In Figure 10 the part of the ellipse between the points B and A is represented with a continuos line, while the portion at the left of B is dotted. Along the stress paths at constant values of p and s which intersect YS in the latter zone, the soil shows a dilatant behaviour with a peak value in

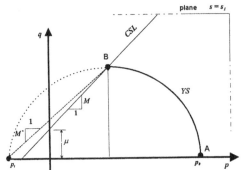

Figure 10. Section of YS with a constant suction plane.

the stress strain relationship. For decreasing values of mean stress, the point representative of the peak stress follow a curve starting from point B and located above the critical state line; in this region the failure locus, in absence of further information, can be assumed to correspond to the dotted part of the ellipse in Figure 10.

The described model is just an extension of Modified Cam Clay to unsaturated soils. For saturated soils, a number of more sophisticated constitutive models have been developed by different authors. In these models the surface which represents the boundary of elastic domain in Modified Cam Clay has the meaning of a State Boundary Surface (SBS) (Atkinson & Sallfors, 1991, Burghignoli et al., 1991). According to these models soil behaviour along stress paths developing inside the SBS can be linear elastic, non linear elastic, or elasto-plastic. Wheeler & Sivakumar (1993, 1995) proposed to extend the idea of SBS to unsaturated soils; going through this assumption the previously mentioned YS in the $\{p, s, q, v\}$ space should be considered as an SBS.

Some more remarks on the relationship between stress obliquity η $(= q/p)$ and dilatancy d $(= \delta\varepsilon_v^p / \delta\varepsilon_s^p)$ are needed to complete the model description. This relationship can be expressed in a very simple form introducing the following modified stress variables (di Prisco et al. 1992, Gens & Nova 1993):

$$p^* = p + |p_t|$$
$$\eta^* = q/p^*$$

(2)

Where p_t represents the value of the mean net stress corresponding to the intersection of YC with negative p axis. The p_t value obviously depends on suction, therefore the introduced variables can be used only to describe constant suction stress paths. Thus the stress obliquity - dilatancy relationship takes the form (in the mentioned hypothesis of associated flow rule):

$$d = \frac{M^{*2} - \eta^{*2}}{2 \cdot \eta^*}$$

(3)

Equation (3) is similar to the stress obliquity - dilatancy relationship of Modified Cam-Clay. It must be noticed that M^* doesn't coincide with the slope M of critical state line. M^* is a function of stress state and it can be shown that:

$$M^* = M^* \left(\frac{\mu}{|p_t|}, \frac{|p_t|}{p^*}, \eta^* \right)$$

$$\therefore M^{*3} + \left[\left(2 \cdot \frac{|p_t|}{p^*} - 1 \right) \cdot M - 2 \cdot \frac{\mu}{|p_t|} \cdot \frac{|p_t|}{p^*} \right] \cdot M^{*2} + \eta^{*2} \cdot M^* - \eta^{*2} \cdot M = 0$$

(4)

It is worth noting that the ratio $\mu / |p_t|$ should vary between 0 and M. The first limiting condition corresponds to the absence of effect of suction on critical state conditions. On the other hand if:

$$\frac{\mu}{|p_t|} = M \tag{5}$$

equations (4) gives:

$$M^* = M \tag{6}$$

which is the maximum possible value for M^*.

In the hypothesis of validity of equation (5) and therefore of equation (6), the stress obliquity - dilatancy relationship of Wheeler & Sivakumar model becomes identical to the ones of Alonso et al. (1987). If $p \geq 0$ the ratio $|p_t| / p^*$ varies between 0 and 1. It is worth noting that if:

$$|p_t|/p^* \to 0 \tag{7}$$

it results:

$$M^* \to M$$
$$\eta^* \to \eta \tag{8}$$

even if the suction value is bigger than zero. Hence expression (3) coincides with the stress obliquity-dilatancy relationship of Modified Cam-Clay:

$$d = \frac{M^2 - \eta^2}{2 \cdot \eta} \tag{9}$$

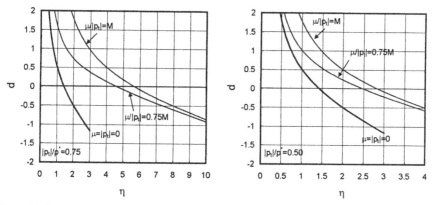

Figure 11. Stress obliquity-dilatancy relationship at different stress levels.

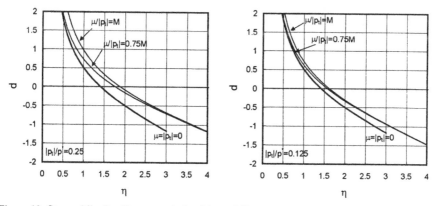

Figure 12. Stress obliquity-dilatancy relationship at different stress levels.

Table 4. Analogy between triaxial test and direct shear test.

Triaxial conditions	Direct shear
mean net stress: p	vertical net stress: $\sigma_v - u_a$
deviatoric stress: q	shear stress: τ
volumetric strain: ε_v	vertical displacement of shear box: dy
deviatoric strain: ε_s	horizontal displacement of shear box: dx
stress path: p = cost	shearing: $\sigma_v - u_a$ = cost

In conclusion equation (3) turns into equation (9) either in the case of nil suction (hence for $|p_t| = 0$) either in the case of "high stress levels" (i.e. $p^* >> |p_t|$).

The stress obliquity-dilatancy relationship of the Wheeler & Sivakumar model is represented in the four diagrams of Figures 11, 12 corresponding to four different stress levels (i.e. $|p_t|/p^*$ values) in the hypothesis of M = 1.418 (i.e. $\phi' = 35°$). In each diagram the relationship is plotted for saturated material ($\mu = 0$ and $|p_t| = 0$) and for two different values of the ratio $\mu / |p_t|$ (M and 0.75·M).

4.2 Model prediction upon wetting

Using the Wheeler & Sivakumar model is possible to simulate the mechanical behaviour of unsaturated soil in triaxial conditions. In order to analyse the results of direct shear tests using the same model, the analogy summarized in Table 4 is assumed.

It must pointed out that the proposed analogy is used in the following only to investigate the adequacy of the described theoretical framework in reproducing behaviour similar to the observed one. Using this analogy a direct shear test can be assimilated to a triaxial test in which shear phase is performed at constant mean net stress. The analyses are focused on the response of unsaturated specimens wetted under a certain applied deviatoric stress. Two different conditions have been analyzed: wetting in strain control conditions and wetting in load control conditions. These two conditions arise when deviatoric strain or deviatoric stress is maintained constant (i.e. respectively $\delta\varepsilon_s = 0$ or $\delta q = 0$) during suction reduction stage.

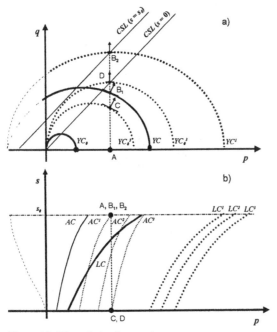

Figure 13. YS evolution in case 1.

186

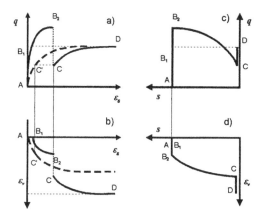

Figure 14. Stress-strain relationships in case 1.

4.2.1 *Strain controlled condition*

As previously mentioned in §3.2 the experimental results show some influence of stress level on mechanical response of material A upon wetting. In order to take into account such observation the analysis of suction reduction process has been subdivided in two different cases.

4.2.1.1 Case 1: "high stress level"

The stress path corresponding to this case is shown in Figure 13 as projection on a constant suction plane. Curves YC_0 and YC represent the sections of YS respectively with the nil suction plane ($s = 0$) and with the initial suction plane (i.e. $s = s_0$). The intersections of YC_0 and YC with p axis represent the yielding isotropic stresses at nil suction, $p_0(0)$ and at initial suction, $p_0(s_0)$. In Figure 13 point A represents the stress state at the beginning of shearing. It's useful to distinguish the following subsequent phases (Figs. 13, 14).

Phase A-B_1: starting from the initial isotropic conditions (point A), deviatoric stress is applied; stress path is initially located inside YC, hence only elastic strains can be developed. In particular, stress-strain relationship is linear and volumetric strains are zero, because mean stresses and suction are constant. However in real soil even inside YC the response is not perfectly elastic and linear.

Phase B_1-B_2: in point B_1 stress path approaches YC in the zone of yielding curve in which soil behaviour is contractant ($d > 0$); with increasing shear strains the sample is subjected to unrecoverable reductions of volume till approaching critical state (point B_2); during the process yielding curve YC changes into YC^1 and YC_0 into YC_0^1 (not represented in figure for sake of simplicity).

Phase B_2-C: during this phase soil is wetted and suction decreases suddenly; while the stress point moves from $s = s_0$ plane to $s = 0$ plane, deviatoric stress decreases abruptly reaching YC_0^1 yielding curve at nil suction. During saturation positive volumetric strains are originated and hence YC_0^1 gets into YC_0^2; it's just on YC_0^2 curve that point C, representative of stress state at the end of saturation, is located. In Figure 14 stress-strain relationship, during saturation, is represented as a sudden decrease of deviatoric stress, to which corresponds an increase of ε_s; specific volume, during saturation, suffers a rapid decrease in agreement with flow conditions that implies a contractant behaviour ($d > 0$).

Phase C-D: in this last phase soil behaviour is still contractant. Saturation, however, has caused a movement of stress point from the intersection between yielding and critical state curves; for this reason dilatancy d is initially greater than the one developed at the end of phase B_1-B_2. With increasing shear strains, soil tends to reach critical conditions (point D); during this process YC_0^2 yielding curve turns into YC_0^3 and deviatoric stress increases asymptotically towards critical value. It must be noticed that the value reached by deviatoric stress in final conditions (point D) is, however, smaller than the one corresponding to point B_2, representative of critical state at $s = s_0$.

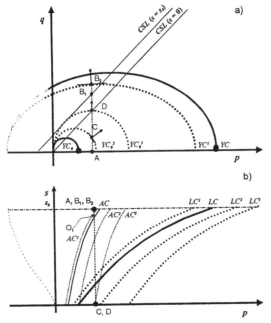

Figure 15. YS evolution in case 2.

The model response shown in Figure 15 is perfectly equivalent to the behaviour observed during direct shear tests performed at high levels of vertical stress. During such tests soil is contractant ($d > 0$) till reaching the maximum value of strength, corresponding to minimum of volumetric strains increment (i.e. $d \rightarrow 0+$). When saturation ends and shear process starts again the value of volumetric strains increment is bigger than the one at the beginning of wetting phase.

4.2.1.2 Case 2: "low stress level"

In the second case, stress path intersects the envelope of peak stress conditions; in fact it intersects YC in a point located at the left side of point B (see Fig. 15). Point A, in Figure 15, indicates stress state at the beginning of shear phase. Even in this case it's useful to distinguish the following subsequent phases (Figs. 15, 16).

Phase A-B_1: starting from isotropic conditions (point A) deviatoric stress is applied; stress path is initially inside YC and hence soil behaviour is linear elastic.

Phase B_1-B_2: in point B_1 stress path goes through CSL inside YC. However real soils can show anelastic behaviour even inside the YC. In Figure 16b a dotted curve (B_1-B'_2) represents volumetric strains occurring in the hypothesis that analyzed material shows anelastic dilatant behaviour when stress state passes above the CSL. In point B_2 peak strength and minimum value of dilatancy d are reached; during the process negative volumetric plastic strains are developed, YC turns into YC^1 and YC_0 turns into YC_0^1.

Phase B_2-C: during this third phase soil is saturated and suction rapidly decreases; while the point representative of stress state moves from $s = 0$ plane to $s = s_0$ plane; deviatoric stress decreases abruptly and reaches the yielding curve at zero suction YC_0^1. During saturation positive volumetric strains are developed and hence YC_0^1 turns into YC_0^2. At the end of saturation point C representative of stress state is located on YC_0^2. In Figure 16 this phase is represented in ε_s-q and ε_s-ε_v planes by means of dotted lines, indicating the rapid decay of deviatoric stress and collapse.

Phase C-D: during this last phase stress point is located at the right side of the intersection between yielding curve and critical state line; thus soil behaviour after saturation becomes contractant ($d > 0$). With increasing shear strains, soil sample suffers an unrecoverable reduction of volume, till reaching critical state conditions (point C). During the process YC_0^2 turns into YC_0^3 and deviatoric stress increases asymptotically towards critical value.

188

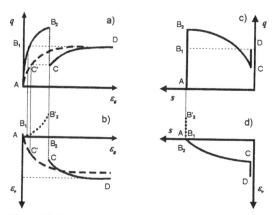

Figure 16. Stress-strain relationships in case 1.

Even in this second case the model response (Fig. 16) is perfectly equivalent to the behaviour observed during direct shear tests performed at lower stress levels. During such tests soil is dilatant ($d < 0$) till reaching the maximum value of strength, corresponding to minimum of volumetric strains increment. When saturation ends and shear process starts again the value of volumetric strains increment is positive ($d > 0$) and soil shows a contractant behaviour till reaching stationary conditions (i.e. $d \rightarrow 0+$).

In conclusion the upper mentioned model can reproduce behaviour of soil upon wetting in triaxial conditions quite similar to those observed in direct shear tests. The model offers even a suggestive interpretation of the origin of the difference between the two analysed cases. In Figures 13b and 15b the projections of the two stress paths on isotropic plane are drawn. In both Figures also LC and the projection AC of the curve in the $\{p, s, q\}$ stress space passing through the yielding ellipse's apex are represented. The latter curve, in the hypothesis of associated flow rule, divides the projection of YS in isotropic plane into two parts: in the part at the right side of AC, soil behaviour is dilatant ($d < 0$), in the part at the left side of AC, soil behaviour is contractant ($d > 0$). Obviously both LC and AC are variable in stress space, due to plastic strains which produce expansions and contractions of YS. In Figures 13b and 15b the subsequent positions of the two curves with increasing strains are reported. In the first case the projection in isotropic plane of the stress point is always located at the right side of the current AC; it means that soil behaviour is always contractant. In the second case this projection is located, at the beginning, at the left side of the current AC (points B_1, B_2 and positions AC, AC_1) and soil behaviour is dilatant; subsequently, during the first part of saturation, stress path goes through the yielding ellipse's apex, and soil behaviour becomes contractant. Transition in isotropic plane (Fig. 15b) is represented by a point with a value of suction higher then the one of point O_1 belonging to AC^1. During saturation, the part of stress path which belongs to the dilatant zone is smaller then the one which develops in the contractant zone, and for this reason during saturation only positive volumetric strains can be observed.

Those considerations can be further clarified by the representation of the wetting processes in the plane η-d. In Figure 17a such representation is reported.

The wetting process at high stress level is qualitatively represented by curve 1-2. The soil state is represented by point 1 at the beginning and by point 2 at the end of saturation. The two points hold to the two stress obliquity-dilatancy curves corresponding respectively to initial suction and nil suction. Saturation causes either a stress obliquity reduction either a dilatancy increase.

The wetting process at low stress level is instead represented by curve 4-5-3. Even in this case saturation causes either a stress obliquity reduction either a dilatancy increase. Thus dilatancy d passes from negative to positive value (soil passes from dilatant to contractant behaviour). It should be even possible that at lower stress level wetting process increases dilatancy without changing its sign (path 6-7 of Fig. 17a). In every cases further shearing after saturation push the points representative of soil state along the stress obliquity-dilatancy curve of saturated material towards critical state conditions (point M).

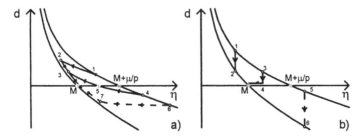

Figure 17. Wetting processes: a) strain controlled conditions; b) load control conditions.

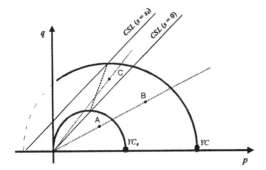

Figure 18. Initial state of wetting processes in load controlled conditions.

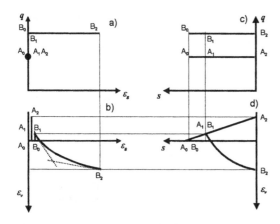

Figure 19. Stress-strain relationships

4.2.2 Load controlled condition

The described model can be used even to predict soil response in load controlled conditions. Three different conditions at the beginning of wetting process have been analyzed corresponding to point A, B and C of Figure 18.

Case A: in this case at the beginning of saturation stress state projection on $s = 0$ plane is located inside YC_0; the whole wetting path develops inside YS; therefore soil experiences only elastic volumetric strains (see points A_0, A_1 and A_2 in Fig. 19).

Case B: in this case at the beginning of saturation stress state projection on $s = 0$ plane is located inside YC but outside YC_0; only the first part of wetting path develops inside YS. During

190

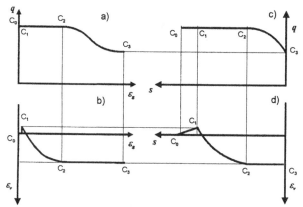

Figure 20. Stress-strain relationship.

this initial phase soil experiences only elastic volumetric strains (see points B_0 and B_1 in Fig. 20). In the second phase of saturation positive volumetric plastic strains develops; in fact if deviatoric stress q is kept constant, an hardening process takes place to maintain stress state inside YS. Therefore according to flow rule during saturation even plastic shear strains develop, as shown in Figure 19a, b (points B_1 and B_2). Path 1-2 of Figure 17b represents in η-d plane the described wetting process; it's worth noting that during saturation positive dilatancy d reduces.

Case C: in this case applied deviatoric stress q_c at the beginning of saturation is larger than deviatoric stress corresponding to applied isotropic stress in nil suction critical state condition. Therefore soil will show a brittle failure if q_c is kept constant till complete saturation. During wetting process a plastic phase develops after an elastic one (points C_0 and C_1 in Fig. 20). The plastic phase can be subdivided in two parts (C_1-C_2 and C_2-C_3 in Fig. 20): initially strain hardening maintain stress state on YS till reaching critical state condition (point C_2); hence further reduction of suction cannot produce further plastic volumetric strain and hardening; therefore saturation must precede in strain control condition (C_2-C_3 in Fig. 20). During this final stage plastic shear strains are virtually infinite while volumetric strains are only elastic and negative. Path 3-4-M of Figure 17a represents in η-d plane the described wetting process; it's worth noting that even in this case positive dilatancy d reduces during saturation. Obviously a wetting processes similar to the one represented by path 5-6 in Figure 17b cannot occur in reality.

5 CONCLUSIONS

As conclusion some remarks on the relevance of analyzed stress paths can be stated. The simple case of an element of unsaturated pyroclastic soil at depth z from the surface of an infinite slope of angle β can probably clarify this point. Consideration of equilibrium of the forces acting on the element allows to evaluate the total normal stress and shear stress acting on a plane at depth z parallel to the slope. These total stresses don't change in the simple hypothesis that saturation occurs in vertical direction along the whole slope (without taking into account the increment of unit weight due to saturation). Therefore an element of soil experiences a wetting process in load control condition. Obviously the slope can remain stable only if applied shear stress is lower than the saturated material shear strength. Conversely total shear stress varies if saturation occurs only in a portion of the slope. In this case elements of soil in the saturated part experience wetting condition intermediate between load and strain controlled. In particular soils elements at the border of the zone in which saturation occurs are in conditions very similar to strain controlled. Saturated portion of slope transmits loads to unsaturated portion due to the wetting process; slope stands stable if unsaturated portion is strong enough to absorb these additional loads.

Two considerations arise from this simple example. The wetting paths analyzed in the paper seem to be extremely significant in some slope stability problems. The predicted response of unsaturated pyroclastic soils upon wetting in load controlled conditions (see Fig.19) shows that even for stress states far from saturated material strength envelope (see point B of Fig. 18) significant plastic shear strains can develops during saturation. Therefore further analyses are needed to asses if these strains can induce unacceptable displacements of the slope. As matter of fact it seems that conventional safety factor evaluations in terms of total stresses could be unsafe in the case of unsaturated pyroclastic slopes.

ACKNOWLEDGEMENTS

The author is indebted to professor S. Aversa for suggesting improvements to the paper, to professor A. Evangelista for co-ordinating the research project ant for defining experimental techniques and to professor A. Pellegrino for defining the research guidelines and for finding funds for experimental program. Dr. A. Scotto di Santolo is thanked for collecting database on pyroclastic material of Sorrentina Peninsula and Dr M. Ramondini for his enthusiastic contribute to in situ activity.

REFERENCES

Alonso, E.E., A. Gens & D.W. Hight 1987. Special problem soils. General report. *IX ECSMFE, Dublin,* 3:1087-1146.

Atkinson, J.H. & G. Sallfors 1991. Experimental determination of stress-strain-time characteristics in laboratory. *Proc. of X ECSMFE, Firenze,* 3: 915-956. Rotterdam: Balkema

Aversa, S., A. Evangelista & M. V. Nicotera 1998. Experimental determination of characteristic curves and permeability functions of partially saturated pyroclastic soil. *Proc. of 2nd Int. Conf. on Unsaturated Soils,* Beijing, 1:13-18.

Burghignoli, A., V. Pane , L. Cavalera, C. Sagaseta, V. Cuellar & M. Pastor 1991. Modelling stress-strain-time behaviour of natural soils. *Proc. of X ECSMFE, Firenze,* 3: 961-979. Rotterdam: Balkema

Cole, P. D. & C. Scarpati 1993. A facies interpretation of the eruption and emplacement mechanism of the upper part of the Neapolitan Yellow Tuff, Campi Flegrei, Southern Italy. *Bull. Volcanology* 55: 311-326.

di Prisco, C., R. Matiotti & R. Nova 1992. A mathematical model of grouted sand behaviour. *Proc. 4th Int. Symp. Numer. Models Geomech., Swansea,* 25-35.

Gens, A. & R. Nova 1993. Conceptual bases for constitutive model for bonded soils and week rocks. *Proc. of Int. Symp. on Geotechnical Engineering of Hard Soils-Soft Rocks, Athens, 20-23 September 1993:* 485-494. Rotterdam: Balkema.

Fredlund, D.G., N.R. Morgenstern & R.A. Widger 1978. The shear strength of unsaturated soils. *Can. Geo. J.* 15:313-321.

Nicotera, M. V. 1998. *Effetti del grado di saturazione sul comportamento meccanico di una pozzolana del napoletano.* Ph.D. Thesis, University of Naples, Italy.

Nicotera, M.V. & S. Aversa 1999. Un laboratorio per la caratterizzazione fisico-meccanica di terreni piroclastici non saturi. *Atti del XX Convegno Nazionale di Geotecnica, Parma, 22-25 settembre 1999,* 201-212. Bologna: Pàtron

Nicotera, M.V., A. Evangelista & S. Aversa 1999a. Determinazione della curva caratteristica e delle funzioni di permeabilità di una pozzolana non satura. *Atti del XX Convegno Nazionale di Geotecnica, Parma, 22-25 settembre 1999,* 213-221. Bologna: Pàtron.

Nicotera, M.V., A. Evangelista & S. Aversa 1999b. Osservazioni sperimentali sulla risposta meccanica di una pozzolana in condizioni lontane dalla saturazione. *Atti del XX Convegno Nazionale di Geotecnica, Parma, 22-25 settembre 1999,* 223-230. Bologna: Pàtron.

Pellegrino, A. 1967. Proprietà fisico-meccaniche dei terreni vulcanici del napoletano. *Atti dell'VIII Convegno di Geotecnica, Cagliari, 6-7 febbraio 1967,* 3: 113-145. Napoli: ESI.

Scarpati C., P.D. Cole & A. Perrotta 1993. The Neapolitan Yellow Tuff – A large volume multiphase eruption from Campi Flegrei, Sothern Italy. *Bull. Volcanology* 55: 343-356.

Scotto di Santolo, A. 2000. *Analisi geotecnica dei fenomeni franosi nelle coltri piroclastiche della Provincia di Napoli.* Ph.D. Thesis, University of Naples, 2000.

Scotto di Santolo, A., M.V. Nicotera, A. Evangelista, A. Pellegrino, M. Ramondini & G. Urciuoli in press. Some remarks on the shear strength of neapolitan pyroclastic deposits. *Proc. of GeoEng 2000, Melbourne, November 2000.*

Wheeler, S.J. & V. Sivakumar 1995. An elasto-plastic critical state framework for unsaturated soil. *Geotechnique* 45(1): 35-53.

Miscellaneous

Author addresses

Mauricio Barrera
mauricio.barrera@upc.es

Departament d'Enginyeria del Terreny, Cartogràfica i Geofísica
Universitat Politècnica de Catalunya
Jordi Girona, 1-3, mòdul D2, Campus Nord
08034 Barcelona, Spain

Marc Buisson
m.buisson@terrasol.com

c/o Prof. S. Wheeler
Glasgow University, Department of Civil Engineering
Rankine Building, Oakfield Avenue
G12 8LT Glasgow, UK
email: Wheeler@civil.gla.ac.uk

Alessandra Di Mariano
alessandra.dimariano@upc.es

Departament d'Enginyeria del Terreny, Cartogràfica i Geofísica
Universitat Politècnica de Catalunya
Jordi Girona, 1-3, mòdul D2, Campus Nord
08034 Barcelona, Spain

Françoise Geiser
f_geiser@hotmail.com

c/o Dr. Lyesse Laloui
Ecole Polytechnique Fédérale de Lausanne
Laboratoire de Mécanique des Sols, DGC - EPFL
1015 Lausanne, Switzerland
email: Laloui@epfl.ch

Cristina Jommi
jommi@stru.polimi.it

Dipartimento di Ingegneria Strutturale
Politecnico di Milano
piazza Leonardo da Vinci 32
20133 Milano, Italy

Marco Valerio Nicotera
nicotera@cds.unina.it

Dipartimento di Ingegneria Geotecnica
Universita' di Napoli Federico II
Via Claudio 21
80125Napoli, Italy

Enrique E. Romero Morales
enrique.romero-morales@upc.es

Departament d'Enginyeria del Terreny, Cartogràfica i Geofisica
Universitat Politècnica de Catalunya
Jordi Girona, 1-3, mòdul D2, Campus Nord
08034 Barcelona, Spain

John Shevelan
cip96js@sheffield.ac.uk

c/o Dr. C.Hird
Department of Civil & Structural Engineering
University of Sheffield
S1 3JD Sheffield, UK
email: c.hird@sheffield.ac.uk

Alessandro Tarantino
tarantin@ing.unitn.it

Dipartimento di Ingegneria Meccanica e Strutturale
Università degli Studi di Trento
via Mesiano 77
38050 Trento, Italy

Roberto Vassallo
rvassall@unina.it

Dipartimento di Ingegneria Geotecnica
Università di Napoli Federico II
Via Claudio, 21
80125 Napoli, Italy

Jean Vaunat
jean.vaunat@upc.es

Departament d'Enginyeria del Terreny, Cartogràfica i Geofisica
Universitat Politècnica de Catalunya
Jordi Girona, 1-3, mòdul D2, Campus Nord
08034 Barcelona, Spain

Mourad Yahia-Aissa
m.yahia-aissa@terrasol.com

c/o Prof. P. Delage
CERMES-ENPC (Ecole Nationale des Ponts et Chaussées)
6 et 8, Av. Blaise Pascal
Cité Descartes, Champs-sur-Marne
77455 Marne la Vallée Cedex 2, France
E-mail: delage@cermes.enpc.fr

Experimental Evidence and Theoretical Approaches in Unsaturated Soils, Tarantino & Mancuso (eds)
© *2000 Taylor & Francis, ISBN 90 5809 186 4*

List of participants

Camillo Airò	Università di Palermo, Italy
Stefano Aversa	Università di Pisa, Italy
Mustafa Aytekin	Karadeniz Technical University, Turkey
Mauricio Barrera	Universitat Politècnica de Catalunya, Spain
Giovanni Bosco	Università di Trento, Italy
Frederic Bourgeois	ANDRA, France
Marc Buisson	University of Glasgow, UK
Francesco Cafaro	Politecnico di Bari, Italy
Carlo Callari	Università di Roma "Tor Vergata", Italy
Manuela Cecconi	Università di Perugia, Italy
Paolo Croce	Università di Cassino, Italy
Stefano Dapporto	Università di Firenze, Italy
Vincenzo De Gennaro	Ecole Nationale des Ponts et Chaussées, France
Pierre Delage	Ecole Nationale des Ponts et Chaussées, France
Alessandra Di Mariano	Università di Palermo, Italy
Carmine Fallico	Università della Calabria, Italy
Settimio Ferlisi	Università di Roma "Tor Vergata", Italy
Walter Ferrazza	Università di Trento, Italy
Alessandro Gajo	Università di Trento, Italy
Francoise Geiser	Ecole Polytechnique Fédérale de Lausanne, Switzerland
Chiara Guiducci	Normale Pisa - ENI, Italy
Cristina Jommi	Politecnico di Milano, Italy
Dimitrios Kolymbas	University of Innsbruck, Austria
Petr Koran	Charles University of Prague, Czech Republic
Sarah M. Springman	Swiss Federal Institute of Technology Zurich, Switzerland
Claudio Mancuso	Università di Napoli Federico II, Italy
Duilio Marcial	Universidad Central de Venezuela, Venezuela
Claudio Matà	IFP (Institut Français du Pétrole), France

John McDougall	Napier University, UK
Roberto Meriggi	Università di Udine, Italy
Giuseppe Modoni	Università di Cassino, Italy
Luigi Mongiovì	Università di Trento, Italy
Marco Valerio Nicotera	Università di Napoli Federico II, Italy
Alessandra Nocilla	Università di Trento, Italy
Roberto Nova	Politecnico di Milano, Italy
Luca Pagano	Università di Napoli Federico II, Italy
Vincenzo Pane	Università di Perugia, Italy
Antonio Pellegrino	Eni-Divisione Agip, Italy
Sebastiano Perisi	Università di Trento, Italy
Nicoletta Peroni	Università di Ancona, Italy
Enrique Romero	Universitat Politècnica de Catalunya, Spain
Giacomo Russo	Università di Cassino, Italy
Alessandra Sciotti	Università di Roma "La Sapienza", Italy
Anna Scotto di Santolo	Università di Napoli Federico II, Italy
John Shevelan	University of Sheffield, UK
Lucia Simeoni	Università di Trento, Italy
Vinayagamoorthy Sivakumar	Queen's University of Belfast, UK
Giuseppe Sorbino	Università di Salerno, Italy
Claudio Tamagnini	Università di Perugia, Italy
Alessandro Tarantino	Università di Trento, Italy
Philipp Teysseire	Swiss Federal Institute of Technology Zurich, Switzerland
Roberto Vassallo	Università di Napoli Federico II, Italy
Jean Vaunat	Universitat Politècnica de Catalunya, Spain

Experimental Evidence and Theoretical Approaches in Unsaturated Soils, Tarantino & Mancuso (eds)
© *2000 Taylor & Francis, ISBN 90 5809 186 4*

Author index

Printed and bound by CPI Group (UK) Ltd, Croydon, CR0 4YY

23/10/2024

01777679-0013